Programmable Controllers

Principles and Applications

John W. Webb

Northcentral Technical College
Wausau, Wisconsin

Merrill Publishing Company
A Bell & Howell Information Company
Columbus   Toronto   London   Melbourne

Cover Photo: Jack Reichert. The photograph shows an unusual application of a programmable controller—the automatic control of a hydroelectric plant. The plant is owned by the Owens Illinois Paper Mill and located in Tomahawk, Wisconsin. The automatic control system includes such perameters as remote operation, electrical load sensing, and available water flow. A Gould Modicon Programmable controller is used for system management control. The engineering of this automatic hydroelectric control system is by L and S Electric of Schofield, Wis. Photo courtesy of L and S Electric.

Published by Merrill Publishing Company
A Bell & Howell Information Company
Columbus, Ohio 43216

This book was set in Century Schoolbook and Helvetica.

Administrative Editor: Stephen Helba
Production Coordinator and Text Designer: Jeffrey Putnam
Art Coordinator: Mark D. Garrett
Cover Designer: Cathy Watterson

Library of Congress Catalog Card Number: 87-61582
International Standard Book Number: 0-675-20452-6
Printed in the United States of America
  2 3 4 5 6 7 8 9 — 92 91 90 89

MERRILL'S INTERNATIONAL SERIES IN ELECTRICAL AND ELECTRONICS TECHNOLOGY

MERRILL'S SERIES IN
MECHANICAL AND CIVIL TECHNOLOGY

Contents

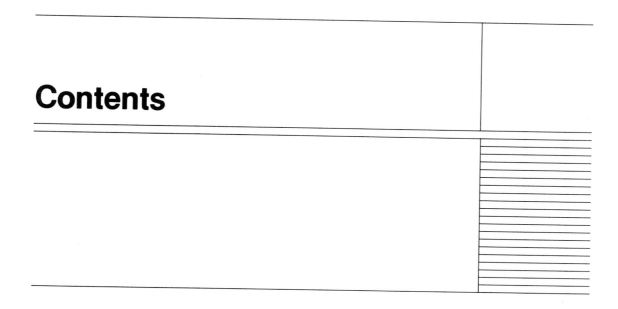

Preface

TRANSITIONAL Functions
SELECT CONTINUOUSLY
ASCENDING SORT
TRANSMIT PRINT
FIRST IN/FIRST OUT and FIRST IN/LAST OUT
PROGRAM and WATCHDOG TIMER CONTROLS
LOOP CONTROL

Appendixes

Preface

The market for programmable controllers is projected to grow to hundreds of millions of dollars a year in the United States. The need for training in PC application is increasing at all levels, in technical schools, colleges, and industry. The purpose of this text is to provide PC training in all of these educational areas.

Each chapter includes learning objectives, an introduction, explanations and examples, and questions. There is a glossary and bibliography at the end. A solutions manual, which includes answers to all chapter questions, is available. Unlike manufacturers' manuals and most PC texts, this book includes many programming examples and exercise problems for each type of PC function. It is also generic, to apply to many different PC models.

Another feature of this book is that it is possible to use a portion of each chapter, depending on the depth of training required. For example, chapter 18 covers the PC Move functions; for a course in basic functions, only Section A on the Move function need be covered. Chapter questions are arranged so that only the applicable sections (the first series of questions) need be used when a function does not have to be covered in depth. For more extensive training on a function, the other portions of a chapter may be covered as required.

For a longer course, all chapters in the text could be covered. For a shorter course, only the following chapters would be essential: 1, 2, 6, 7, 8, 11, 12, 13, 14, 15, 16, 17, 18, 19, 20, and 23. Depending on students' backgrounds, other chapters could be included or omitted. In some cases the sequential order of chapters may not follow the order in the book.

The text is divided into six parts: *One,* PC Equipment and Hardware; *Two,* Programs and Software; *Three,* Basic PC Functions; *Four,* Intermediate Functions; *Five,* Functions Involving Individual Register Bits; and *Six,* Advanced Functions. The chart shows which sections would probably be of interest in the listed training situations. Certain sections are of high interest (H) for a given audience, others of medium interest (M), and some of low interest (L). Of course, each training situation calls for a varying amount of time and emphasis on each section and chapter. Also, individual chapters can be covered completely or lightly, depending on the students' background.

	Section of this Book					
	1	**2**	**3**	**4**	**5**	**6**
Limited Length PC School	L	M	H	M	M	L
In-house Manufacturer's School	M	H	H	M	L	L
Electrical Trades	H	M	H	H	M	L
Vocational Electrical/Mechanical Programs	M	M	H	M	M	L
Associate Degree, Electronics Related	H	H	H	H	H	M
Associate Degree, Electromechanical/ Robotics	H	H	H	H	H	H
Technology, Four-year School	M	H	H	H	H	H
Engineering School	M	H	H	H	H	H

Acknowledgements

I would like to thank the reviewers for their helpful comments:

Roger Bertrand, Central Maine Technical Institute; Ken Edwards, International Brotherhood of Electrical Workers; Phil Henning, Williamsport Community College; Thomas Kissell, Terra Technical College; E.W. McCullough, Rowan Technical College; William G. Pfautz, De Kalb Area Technical School; Terry S. Taebel; and Stephen Tubbs, Pennsylvania State University—McKeesport. I also want to thank the following companies and people for their support and cooperation in preparing *Programmable Controllers:* Giddings and Lewis Electronics; Westinghouse Electric, Numa Logic; Texas Instruments, Inc.; General Electric; Eaton Corp, Cutler Hammer Products; TII Robotics Systems, Division of TII Robotic Systems, Inc.; North Central Technical Institute; L and S Electric, Schofield, Wisconsin; and Paula Webb Clark.

To my wife, Thelma

SECTION ONE

PC EQUIPMENT AND HARDWARE

Introduction

At the end of this chapter, you will be able to

☐ Discuss the history of the programmable controller.
☐ List and discuss advantages and disadvantages of the PC.
☐ Evaluate knowledge levels needed for PC programming and operating.

In this introductory chapter, we will tell what a programmable controller (PC) is. We will then discuss the evolution of relay logic and computer systems into the present-day programmable controller. We will also list and discuss some advantages and disadvantages of using a PC over other control systems. Finally, the knowledge level required for PC programming and operating will be evaluated.

DEFINITION OF A PROGRAMMABLE CONTROLLER

A programmable controller is a user-friendly electronic computer that carries out control functions of many types and levels of complexity. It can be programmed, controlled, and operated by a person unskilled in operating computers. The programmable controller essentially draws the lines and devices of ladder diagrams. The resulting drawing in the computer takes the place of much of the external wiring required for control of a process. The programmable controller will operate any system that has output devices that go on and off. It can also operate any system with variable outputs. The programmable controller can be operated on the input side by on/off devices or by variable input devices.

EVOLUTION TO THE PRESENT PROGRAMMABLE CONTROLLER

The first PC systems evolved from conventional computers in the late 1960s and early 1970s. These first PCs were mostly installed in automotive plants. Traditionally, the auto plants had to be shut down for up to a month at model changeover time. The early PCs were used along with other new automation techniques to shorten the changeover time. One of the major time-consuming changeover procedures had been the wiring of new or revised relay and control panels. The PC keyboard re-programming procedure replaced the rewiring of a panel full of wires, relays, timers, and other components. The new PCs helped reduce reprogramming time to a matter of a few days.

There was a major problem with these early-1970s computer/PC repro-gramming procedures. The programs were complicated and required a highly trained programmer to make the changes. Through the late 1970s, improvements were made in PC programs to make them somewhat more user friendly; in 1978, the introduction of the microprocessor chip increased computer power for all kinds of automation systems and lowered the computing cost. Robotics, automation devices, and computers of all types, including the PC, consequently underwent many improvements. PC pro-grams became more understandable to more people. PCs became more affordable, as well.

In the 1980s, with more computer power per dollar available, the PC came into exponentially increasing use. Some large electronics and com-puter companies and some diverse corporate electronics divisions found that the PC had become their greatest volume product. The market for PCs grew from a volume of $80 million in 1978, to $1 billion dollar per year in the mid 1980s and is still growing. Even the machine tool industry, where computer numerical controls (CNC) controls have been used in the past, is using PCs. PCs are also used extensively in building energy and security control systems. Other nontraditional uses of PCs, such as in the home and in medical equipment, are increasing in the 1980s.

ADVANTAGES OF THE PROGRAMMABLE CONTROLLER

The following are some of the major advantages of using a programmable controller:

Flexibility. In the past, each different electronically controlled produc-tion machine required its own controller; 15 machines might require 15 different controllers. Now, it is possible to use just one model of a PC to run any one of the 15 machines. Furthermore, you would probably need fewer than 15 controllers, because one PC can easily run many machines.

Each of the 15 machines under PC control would have its own distinct program.

Implementing changes and correcting errors. With a wired relay-type panel, any program alterations require time for rewiring of panels and devices. When a PC program circuit or sequence design change is made, the PC program can be changed from a keyboard sequence in a matter of minutes. No rewiring is required for a PC-controlled system. Also, if a programming error has to be corrected in a PC control ladder diagram, a change can be typed in quickly.

Large quantities of contacts. The PC has a large number of contacts for each coil available in its programming. Suppose that a panel-wired relay has four contacts and all are in use when a design change requiring three more contacts is made. It would mean that time must be taken to procure and install a new relay or relay contact block. Using a PC, however, would only require that three more contacts be typed in. The three contacts would be automatically available in the PC. Indeed, a hundred contacts can be used from one relay—if sufficient computer memory is available.

Lower cost. Increased technology makes it possible to compact more functions into smaller and less expensive packages. In the mid-1980s you can purchase a PC with numerous relays, timers, counters, a sequencer, and other functions for a few hundred dollars.

Pilot running. A PC programmed circuit can be pre-run and evaluated in the office or lab. The program can be typed in, tested, observed, and modified if needed, saving valuable factory time. In contrast, conventional relay systems have been best tested on the factory floor, which can be very time consuming.

Visual observation. A PC circuit's operation can be seen during operation directly on a CRT screen. The operation or mis-operation of a circuit can be observed as it happens. Logic paths light up on the screen as they are energized. Troubleshooting can be done quicker during visual observation.

In advanced PC systems, an operator message can be programmed for each possible malfunction. The malfunction description appears on the screen when the malfunction is detected by the PC logic (for example, "MOTOR #7 IS OVERLOADED"). Advanced PC systems also may have descriptions of the function of each circuit component. For example, input #1 on the diagram could have "CONVEYOR LIMIT SWITCH" on the diagram as a description.

Speed of operation. Relays can take an unacceptable amount of time to actuate. The operational speed for the PC program is very fast. The speed for the PC logic operation is determined by scan time, which is a matter of milliseconds.

Ladder or Boolean programming method. The PC programming can be accomplished in the ladder mode by an electrician or technician. Alternately, a PC programmer who works in digital or Boolean control systems can also easily perform PC programming.

Reliability. Solid state devices are more reliable, in general, than mechanical or electrical relays and timers. The PC is made up of solid state electronic components with very high reliabilities.

Simplicity of ordering control system components. A PC is one device with one delivery date. When the PC arrives, all the counters, relays, and other components also arrive. In designing a relay panel, on the other hand, you may have 20 different relays and timers from 12 different suppliers. Obtaining the parts on time involves various delivery dates and availabilities. With a PC you have one product and one lead time for delivery. In a relay system, forgetting to buy one component would mean delaying the start up of the control system until that component arrives. With the PC, one more relay is always available—providing you ordered a PC with enough extra computing power.

Documentation. An immediate printout of the true PC circuit is available in minutes, if required. There is no need to look for the blueprint of the circuit in remote files. The PC prints out the actual circuit in operation at a given moment. Often, the file prints for relay panels are not properly kept up to date. A PC printout is the circuit at the present time; no wire tracing is needed for verification.

Security. A PC program change cannot be made unless the PC is properly unlocked and programmed. Relay panels tend to undergo undocumented changes. People on late shifts do not always record panel alterations made when the office area is locked up for the night.

Ease of changes by reprogramming. Since the PC can be reprogrammed quickly, mixed production processing can be accomplished. For example, if part B comes down the assembly line while part A is still being processed, a program for part B's processing can be reprogrammed into the production machinery in a matter of seconds.

These 13 items are some of the advantages of using a programmable controller. There will, of course, be other advantages in individual applications and industries.

DISADVANTAGES OF THE PROGRAMMABLE CONTROLLER

Following are some of the disadvantages of, or perhaps precautions for, using PCs:

Newer technology. It is difficult to change some personnel's thinking from ladders and relays to the PC computer concepts.

Fixed program applications. Some applications are single-function applications. It doesn't pay to use a PC that includes multiple programming capabilities if they are not needed. One example is in the use of drum controller/sequencers. Some equipment manufacturers still use a mechanical drum with pegs at an overall cost advantage. Their operational sequence is seldom or never changed, so the reprogramming available with the PC would not be necessary.

Environmental considerations. Certain process environments, such as high heat and vibration, interfere with the electronic devices in PCs, which limits their use.

Fail-safe operation. In relay systems, the stop button electrically disconnects the circuit; if the power fails, the system stops. Furthermore, the relay system does not automatically restart when power is restored. This, of course, can be programmed into the PC; however, in some PC programs, you may have to apply an input voltage to cause a device to stop. These systems are not "fail-safe." This disadvantage can be overcome by adding safety relays to a PC system, as we will see later in this text.

Fixed-circuit operation. If the circuit in operation is never altered, a fixed control system such as a mechanical drum, for example, might be less costly than a PC. The PC is most effective when periodic changes in operation are made.

KNOWLEDGE LEVEL FOR PC PROGRAMMING

A person knowledgeable in relay logic systems can master the major PC functions in a few hours. These functions might include coils, contacts, timers, and counters. The same is true for a person with a digital logic background. For persons unfamiliar with ladder diagrams or digital principles, however, the learning process takes more time.

A person knowledgeable in relay logic can master advanced PC functions in a few days with proper instruction. Company schools and operating manuals are very helpful in mastering these advanced functions. Advanced functions in order of learning might include sequencer/drum controller, register bit use, and move functions.

EXERCISES

1. Discuss the evolution of relay logic and the computer into the PC.
2. List 14 advantages of using a PC. Use items discussed in the text plus some of your own ideas.
3. List six disadvantages of using a PC. Again, use the text ideas plus your own.
4. Based on your own knowledge and skill level in the areas of relay and digital logic, evaluate the level of difficulty you will have in learning PC programming and operations. Do the same for two or three typical factory workers, electricians, or technicians.

Programmable Controller System Description

2

At the end of this chapter, you will be able to

☐ List the five major parts of a PC system.
☐ Describe the function of each of the five parts.
☐ Describe how the parts of the system are connected electrically.
☐ List the major classifications of input/output (I/O) modules.
☐ Outline the major precautions to follow when connecting I/O modules.
☐ Describe the general procedure to follow in using a personal computer and program disk system to program a PC.
☐ Explain baud rates and how they are set.

INTRODUCTION

Chapter 2 will describe the components and modules that make up a PC control system. A simple PC system is housed in one, or possibly two enclosures, each of which would include multiple functions. A more complex PC, controlling a large process, may have three to five or more separate interconnected enclosures containing the PC subsystems.

Illustrations of the various subparts of a PC will be shown, as will general connection paths. The electrical interconnections of the various PC parts will be described in general terms. Details of the connections will be discussed in chapter 4.

Most of the PC electrical connecting is easily done with single cables between units. However, connecting the input and output modules to the outside world can be fairly complicated. Input and output module connections to the processes will be discussed in this chapter and throughout the text. The proper setting of module switches will be described in this chapter.

There are personal computer and operating disk systems available to carry out PC programming. These personal computer systems will be discussed, also. PC systems operate at different computer rates. The rate, commonly called "baud" rate, depends on what parts of the PC system are communicating. A PC may operate at a baud rate of 9600 with the CPU, at 1200 with a tape recorder, and at 2400 when working with a printer. Proper setting and resetting of the rate, which will be discussed later in the chapter, is accomplished by following the PC operating manual instructions.

THE OVERALL SYSTEM

Figure 2–1 shows, in block form, the five major units of a PC system and how they are interconnected. The five major parts, each of which will be described later in detail, are:

- ☑ Central Processing Unit (CPU). The "brain" or heart of the system.
- ☑ Programmer/Monitor (PM). The programmer is the keyboard on which the program instructions are typed by the user. The monitor is a televisionlike screen on which the typed or operating information is displayed.

Figure 2–1
PC System
Layout and
Connection

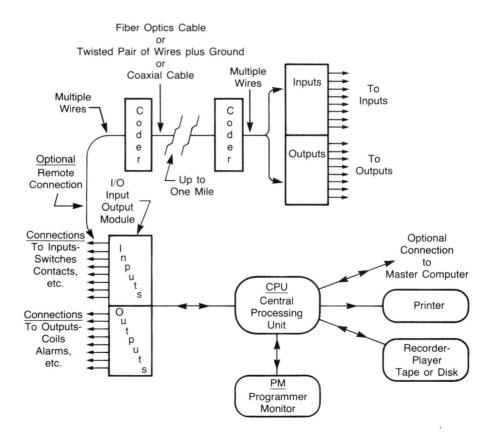

☑ The Input/Output Modules (I/O). The input module has terminals into which the user enters outside process electrical signals. The output module has another set of terminals that send action signals to the process. A remote electronic system for connecting I/O modules to remote locations can be added if needed. The actual operating process under PC control can be thousands of feet from the CPU and its I/O modules.

☑ The Printer. A device on which the program in the Central Processing Unit may be printed out. Additionally, operating information may be printed out upon command.

☑ The Program Recorder/Player. Some PC systems use tape devices; others use floppy disk systems, devices that can externally record the programs in the CPU. The recorded programs, which are then stored, may later be re-entered into the CPU if the original program is lost or develops an error.

For large operations, another possible option is the connection of the CPU to a master computer. The master computer may be used in a large factory or process system to coordinate many individual PC systems. The interconnecting electrical busses are sometimes referred to as data highways.

THE CENTRAL PROCESSING UNIT

The central processing unit (CPU) is the heart of the PC system. A typical central processing unit is shown in figure 2-2. The CPU you use may be smaller or larger than the one shown, depending on the size of the process to be controlled. It is important to size the system CPU according to the internal memory needed to run the process. Controlling a small operation requires only a small PC unit with limited memory; controlling a larger system would require a larger unit with more memory and functions.

Some CPUs can have additional memory easily added on at a later date; others cannot be added to or expanded. Advanced planning with the manufacturer is required to match present and future needs with the size of the system being purchased.

The CPU contains various electrical receptacles for connecting the cables that go to the other PC units. It is important to connect the proper receptacles with the correct cables supplied by the manufacturer.

Many CPUs contain backup batteries that keep the operating program in storage in the event of a plant power failure. Typical retentive backup time is one month. The basic operating program is permanently stored in the CPU and is not lost when input power is lost. However, the process control ladder program is not permanently stored. Battery backup power enables the CPU to retain the operating ladder program in the event of

Figure 2-2
Central Processing Unit (CPU) (Courtesy of Westinghouse)

power loss. Only the operating program can be lost or erased when PC CPU power is lost.

The CPUs all have operational switches, some of which require a key to prevent unauthorized personnel from running a turned-off process. The key-type switch also can prevent unauthorized alterations to the operating program of the system. The switch positions vary from manufacturer to manufacturer, but are similar. Typical positions are:

☐ Off—System cannot be run or programmed.

☐ Run—Allows the system to run but no program alterations can be made.

☐ Disable—Turns all outputs to Off or sets them to the inoperable state.

☐ Monitor—Turns on screen that displays operating information.

☐ Run/Program—System can run, and program modifications can be made to it while it is running. This mode must be used with caution. In this mode, the program cannot be completely erased (for safety) but can only be modified. To delete an entire program, the key must be in the *disable* position.

☐ Off/Program or Program—System cannot run but can be programmed or reprogrammed.

Some manufacturer's programmers may have other special key positions in addition to these.

THE PROGRAMMER/MONITOR

Figure 2–3 shows some typical large programmer/monitors with large cathode ray screens. Figure 2–4 shows some typical, small, hand-held programmers with small display windows. The difference in display size is

Figure 2-3
Large Screen Program/Monitors (Courtesy of Giddings and Lewis and General Electric)

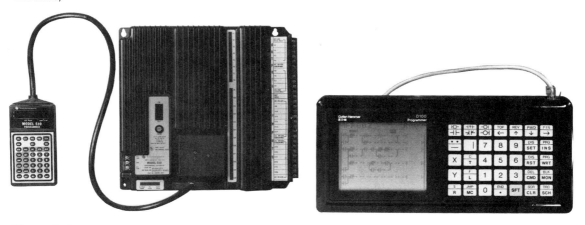

Figure 2-4
Hand-Held Program/Monitors (PM) (Courtesy of Cutler Hammer and Texas Instruments)

directly related to cost. The units in figure 2–3 cost considerably more, but give more information on the screen. A large monitor screen shows an entire circuit of one to five rungs. The smaller hand-held screen display shows only one part of the circuit at a time. With the smaller unit, you will have to go through two or three steps to see all of just one ladder rung.

The Programmer/Monitor (PM) is connected to the CPU by a cable. After the CPU has been programmed, the PM is no longer required for CPU and process operation and can be disconnected and removed. Therefore, you may need only one PM for a number of operational CPUs. The PM may be moved about in the plant as needed. The same PM can even be used in the office or lab to pretest programs.

The PM keyboard and screen operations will be discussed in detail in chapter 4.

INPUT AND OUTPUT MODULES

Now that we have the PC CPU programmed, we get information in and out of the PC through the use of input and output modules. The input module terminals receive signals from wires connected to switches, indicators, and other input information devices. The output module terminals provide output voltages to energize motors and valves, operate indicating devices, and so on.

There are typically 4, 8, 12, or 16 terminals per module. There may be the same number of terminals for a PC's input and output modules, but often there are different numbers of terminals for input and output; for example, a system may have 12 inputs and 8 outputs. A typical module is shown in figure 2–5.

In smaller systems, the input and output terminals may be included on the same frame as the CPU. Figure 2–6 shows two such units.

In other, larger PC systems, the input and output modules are separate units. In these larger systems, modules are placed in groups on racks, as shown in figure 2–7. The racks are connected to the CPU via appropriate connector multiconductor cables.

Figure 2-5
Input and Output Modules (Courtesy of Texas Instruments)

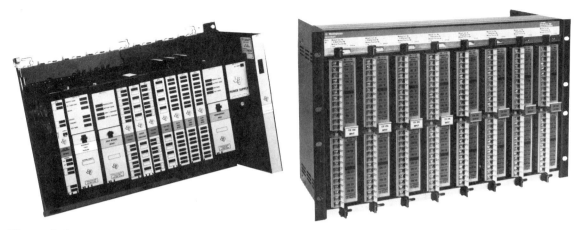

Figure 2-6
I/O Terminals Combined with a CPU (Courtesy of Texas Instruments and
General Electric)

Figure 2-7
Rack Mounts
for I/O Modules
(Courtesy of
Giddings
and Lewis)

Typically, up to 256 terminals may be controlled using only 9 to 24 interconnecting wires. The exact number of wires is determined by the type of computer configuration used for terminal-to-CPU information interchange. The electrical controlling signals from the CPU to the I/O terminals are coded and decoded electronically, making 256 wires for 256 terminals unnecessary.

For multiple modules in a rack, it is necessary to set module switches for each individual module. These settings specify each module's operational number series. Again, for 256 inputs and 256 outputs on a rack, there are 9 to 24 wires in the cable connected to the CPU. Each rack group knows what numbers it should respond to by the system of dual in-line package (DIP) switch settings.

Figure 2–8 shows how these switches appear on a typical module. The settings of these small switches are specified in the manufacturer's manuals. Say, for example, we wish a given 16-terminal module to respond to input signals 17 through 32. The manufacturer says that for the numbers 17 through 32 the eight small DIP switches on the module are to be set at 0-1-0-0-1-1-1-0. Switch 1 is Off (O), 2 is On (1), and so forth. Other modules for other different series of numbers would have correspondingly different switch settings.

One word of caution: some I/O modules have additional internal switches as well as the external visible groups of switches. Check the operations manual to see whether these exist and, if so, how they must be set for proper number response.

Rack Switch Settings and I/O Circuit Reference Numbers.

Rack Switch Setting[1]

Group Select	Top Select	Bottom Select	Responds to CPU Number
1	1	1	1-16
1	2	2	17-32
1	3	3	33-48
1	4	4	49-64
2	1	1	65-80
2	2	2	81-96
2	3	3	97-112
2	4	4	113-128
3	1	1	129-144
3	2	2	145-160
3	3	3	161-176
3	4	4	177-192
4	1	1	193-208
4	2	2	209-224
4	3	3	225-240
4	4	4	241-256

[1]Indicated switch of each set is on while other three switches in set are off.

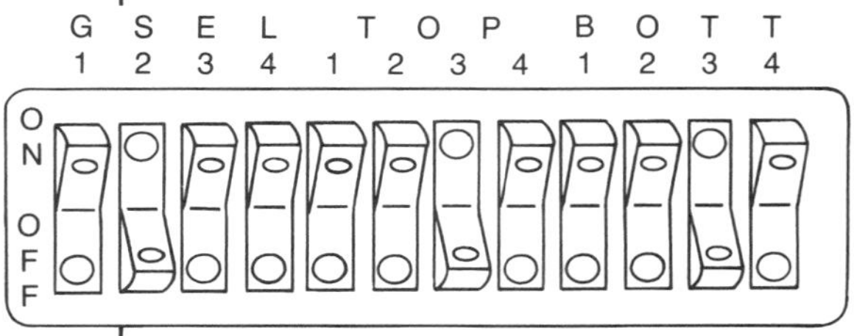

Above Settings = 2/3/3

Note: These settings are for external switches. There may be additional internal switches which also must be set.

Figure 2–8
Number Identification Switch Settings for I/O Modules

If two output or two input modules have the same DIP switch settings, they will act identically. Conversely, if no modules are set for certain input or output groups of numbers, say, for example, 48 through 63, inputs to terminals 48 through 63 are dead-ended electrically at the input module. Their status signal has no path to enter the CPU. The same principle applies to output terminals. Nothing will happen externally when the CPU sends out action signals to terminals for which DIP switches have not been set.

A most important consideration for an I/O module is the module's voltage and current rating. Both voltage and current must match the electrical requirements of the system to which it is connected. An input module rated at 24 volts DC will not work on 120 volts AC and may even be damaged if the module fuse does not act quickly. An output device requiring 4.5 amps cannot be turned on by a 2-amp output module; the module fuse would blow. PC manufacturers have a wide variety of input and output modules available. Module ratings are chosen by the manufacturers to cover the most common applications of their customers. Typical ratings available from manufacturers are shown in figure 2–9.

REMOTE LOCATION TO INPUT/OUTPUT MODULES

Sometimes the processes to be controlled by a PC are a long distance from the CPU or from each other. The normal input and output electrical signals will be reduced to a value too low for module recognition due to long interconnecting wires. Remote amplifier units are available for cases such as these. A typical remote setup is shown in figure 2–10. The input and output signals from the CPU are coded by an adjacent coding unit into digital electrical pulses. The pulses are transmitted over two wires, or by a fiber optics system, to the remote location. At the remote location, a matching station decodes the digital signals. The digital pulses are decoded back into the separate signals that feed the remote modules. The signals originally leaving the CPU are therefore exactly duplicated at the remote modules—a module a mile away will operate as if it were 10 feet away.

DISCRETE AND ANALOG MODULES

The first 22 chapters of this text discuss the most common type of module, the discrete type. Inputs in the discrete type of module are either on or off and the outputs are either energized or deenergized.

Chapter 23 will cover the basic principles of a different type of module—the analog. These analog modules work with variable signals with varying values. Another advanced module type for analog loop control systems will be covered briefly in chapter 24.

Many newer modules have an internal computer for faster process control operation. For example, a process might have a critical input which

VOLTAGE LEVEL	MODULE	CATALOG NUMBER IC600BF	CIRCUIT QUANTITY	UNIT OF I/O LOAD
115 Vac/dc	Input	804	8	2
115 Vac, 2 amp	Output	904	8	9
115 Vac/dc, Isolated	Input	810	6	2
115 Vac, 3½ amp, Isolated	Output	910	6	8
115 Vac, 4 amp, Protected	Output	930	4	8
220 Vac/dc	Input	805	8	2
220 Vac, 2 amp	Output	905	8	9
230 Vac/dc, Isolated	Input	812	6	2
220 Vac, 3½ amp, Isolated	Output	912	6	8
12 Vac/dc	Input	806	8	2
12 Vdc, Sink	Output	906	8	7
12 Vdc, Source	Output	907	8	7
24–48 Vac/dc	Input	802	8	2
24 Vdc, Sink	Output	902	8	7
24 Vdc, Source	Output	908	8	7
48 Vdc, Sink	Output	903	8	7
48 Vdc, Source	Output	909	8	7
120 Vdc, 1½ amp	Output	924	8	5
5 VTTL/10–50Vdc w/o Lights	Input	811	32	4
5 VTTL w/o Lights	Output	911	32	3
10–50 Vdc, Sink, w/o Lights	Output	913	32	3
10–50 Vdc, Source, w/o Lights	Output	919	32	3
5 VTTL/10–50 Vdc with Lights	Input	831	32	4
5 VTTL with Lights	Output	921	32	3
10–50 Vdc, Sink, with Lights	Output	923	32	3
10–50 Vdc, Source, with Lights	Output	929	32	3
100 VA Reeds (NO/NC)	Output	914	6	17
0–10 Vdc Analog	Input	841	8	29
− 10 to + 10 Vdc Analog	Input	842	8	29
4–20ma/1–5Vdc Analog	Input	843	8	29
0–10 Vdc Analog	Output	941	4	29
− 10 to + 10 Vdc Analog	Output	942	4	29
4–20ma Analog	Output	943	4	29
Thermocouple Type J	Input	813	8	9
Thermocouple Type K	Input	814	8	9
Thermocouple Type S	Input	815	8	9
Thermocouple Type T	Input	816	8	9
Axis Position, Type 1	Output	915	1	42
High Speed Counter	I/O	827	1	19
Interrupt	Input	808	8	3
I/O Local Receiver		800		9
I/O Local Transmitter		900		34
I/O Remote Receiver		801		42
I/O Remote Driver		901		38

Figure 2-9
Typical Available Module Ratings (Courtesy of General Electric)

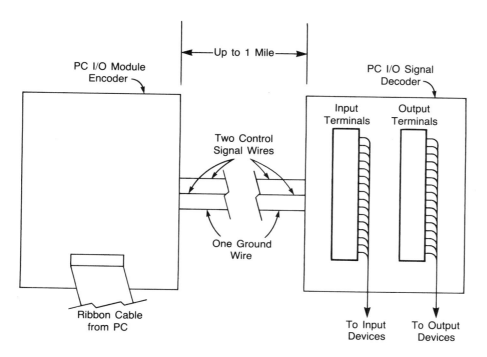

Figure 2-10
Remote PC
Operation

must be acted upon immediately for the safety of process personnel. Sending signals to the CPU, CPU analysis, and return-signal time takes too long. The module can do the analysis continuously and quickly and can take action immediately.

There is one major precaution to be considered with PC output modules. In relay operation, when a relay contact is open, there is no current flow in the associated controlled circuit. However, PC output modules, when turned off, are not strictly off. A small leakage current from the output terminal to the output module still exists, even though the output module is turned off. The output current of each module terminal comes from the output of a thyristor semiconductor called a triac. When not turned on, the triac still puts out a small amount of current. The leakage current is a matter of a few milliamps and is often of no consequence; however, the leakage current may have to be considered in some applications. For example, a PC output terminal might supply a neon bulb which indicates that the output is on; the neon will glow dimly when the module is off due to the leakage current. It might be necessary to add an amplifier or shunting resistors in the electrical output system.

PRINTERS

A typical PC printer is shown in figure 2–11. Printers are used to record information from the CPU for visual analysis. Lengthy ladder programs

Figure 2-11
A PC Printer
(Courtesy of
Cutler Hammer)

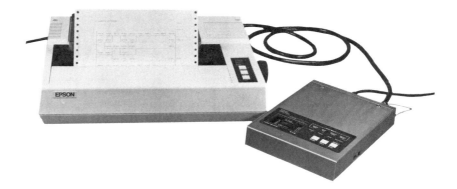

cannot be completely shown on a screen; typically, a screen shows only one to five rungs. A printout on a continuous paper roll can show ladder diagrams and programs of any length. In industrial settings, the complete diagram can be used to analyze the complete circuit. In educational settings, printouts may be used for written assignments to check for correct program construction.

There are many different types of PC information that can be printed out:

☐ Ladder diagrams (which may include coil/contact cross references)
☐ Status of registers
☐ Status and listing of forced conditions
☐ Timing diagrams of contacts
☐ Timing diagrams of registers
☐ Other special diagrams or information

Chapter 14 will describe in detail how to use the printer. It will also list the possible benefits of the various printouts.

PROGRAM RECORDING DEVICES—TAPE OR DISK

Figure 2–12 depicts a typical PC program recorder/player. Any ladder program or other specified information may be recorded on tape from the CPU. Also, any previously recorded program may be quickly put back into the CPU from the tape or disk. The recordings are particularly useful when the program in the CPU is lost or misprogrammed. Playing the tape back into the system is a 1-to-15-minute job. Retyping in a program might take from 15 minutes up to several days, plus time for correcting any mistakes. Another use for recordings is to keep original programs secure (locked up). In addition, if an operating program is altered, a recorded program assures that an original is available for quick reference and comparison.

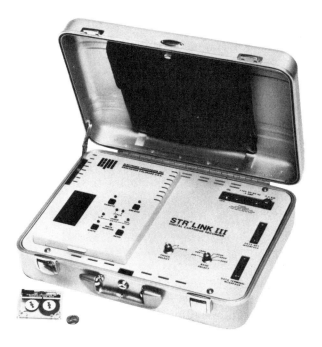

Figure 2-12
PC Tape
Recorder/Player
(Courtesy of
Giddings and
Lewis)

    The PC tape players are not normally inexpensive. They are specialized, high-resolution devices that can cost ten times as much as the usual audio recorder. One advantage of these recording systems is that the system is "user friendly." Operating the recording system is easy since the PC screen in the TAPE mode calls out what to do step by step. Details will follow in chapter 14.

    As in other computer systems, a PC disk drive system has three major advantages over a PC tape drive system: speed, rapid program accessibility, and quantity of data that can be stored.

Figure 2-13
PC Disk
Recorder/
Player
(Courtesy of
General Electric)

BAUD RATE SETTING

Various parts of a PC system require different computer operational rates for proper operation. These rates are called baud rates. A PC CPU computer may "converse" with its keyboard at a rate of 4800 bauds. For remote operation it might use 2400 bauds. Two peripheral devices might use rates of 600 and 1200 bauds. The baud rates vary for each manufacturer and its individual PC device. Each device's baud rate is set automatically when the PC is turned on. The baud rates may have to be reset for certain modes of PC operation. If, for example, you attempt to print a ladder diagram and get an unreadable result, it may be that the baud rate is set incorrectly. Refer to and follow the manufacturer's manual program section on setting peripheral baud rates.

EXERCISES

1. We are replacing seven relays, one timer, and two push-button stations with a PC system. What kind of a system would you purchase and why? What further information might you need to make a purchase decision?

2. A factory has five sections, each with its own process. One of the sections is in a building far away from the others. Programming alterations are required weekly for the processes. What kind of PC system would you recommend, and why? Draw a layout block diagram of the system.

3. Obtain the manuals from one or two different PC models. List the various types of input and output modules available for each model.

4. From the manuals in exercise 3, determine and list the baud rates at which the models operate, including the peripheral baud rates, etc.

Internal Operation of the CPU and I/O Modules

3

At the end of this chapter, you will be able to

☐ List and define the functions of each of the major sections of a PC CPU.
☐ List and describe the operation of the major classes of IC chips used in a PC CPU.
☐ Describe how an AC-in/DC-out power supply functions.
☐ List and describe typical subparts of the PC CPU's alterable memory.
☐ Describe, using diagrams, how a CPU processes information internally.
☐ Describe the operation of a typical input module.
☐ Describe the operation of a typical output module.

INTRODUCTION

The first part of chapter 3 discusses the internal operation of the PC CPU. The CPU is a microprocessor-controlled computer, the operation of which is similar to that of an industrial or personal computer. The CPU's various subsections and their inter-relations will also be described in the first part of the chapter.

The second part of chapter 3 discusses the operation of the input and output modules. The input and output, or I/O, modules are electrical signal converters. The input modules are actuated or energized by process inputs. The input module then converts the input electrical signal to an output signal for CPU analysis. The resulting signal sent to the CPU from the input module becomes a 5-volt digital pulse. The pulses are analyzed by the CPU to determine which inputs are on.

The output module receives digital 5-volt electrical pulses from the CPU. Determination of on-off output status is set from the pulses. The pulse pattern is determined by the CPU ladder logic analysis. The pulses from the CPU are decoded by the output modules. The decoding process results in appropriate electrical output action. The output action, on or off, is then fed to the process being controlled.

THE CENTRAL PROCESSING UNIT OPERATION

Figure 3–1 illustrates the major operational sections of a typical PC CPU. The major sections are the power supply, the fixed memory, the alterable memory, the processor, and the battery backup. The power supply converts the input electrical supply power to electrical power usable by the CPU and the entire computer.

Figure 3–1
Operational Sections of a PC CPU

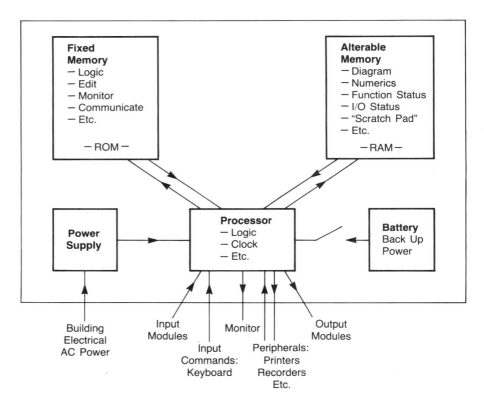

The fixed memory contains the program set by the manufacturer. It is set into special IC chips called Read Only Memory (ROM). The fixed memory in ROM cannot be altered or erased during the CPU's operation. The fixed program memory, often called "non-volatile" memory, is not lost when power is removed from the CPU.

The alterable memory contains many sections which will be outlined later in the chapter. Its information is also stored on IC chips that can be programmed, altered, and erased by the programmer/user. The alterable memory is stored mainly on Random Access Memory (RAM) chips. Information can be written onto or read from a RAM chip. RAM is often called Read/Write memory. The typical RAM chip will lose any information it has stored when input power is lost. It is therefore a "volatile" device; that is, its memory is erased when power is lost. There are some types of RAM chips that are non-volatile; these will be described later in the chapter.

Note that, as stated in chapter 2, there is battery backup power in most CPUs. Power backup prevents the loss of any program that has been inserted into the PC RAM if input power fails.

As illustrated in figure 3–1, the processor section has computer flow connections to other subsections of the CPU and to outside devices. The processor is the controller that keeps information going from one place to another. It contains a clock circuit that allows each transfer of information to take place at the proper time. Details of the processor will be covered later in this chapter.

OTHER IC CHIPS USED IN PC CENTRAL PROCESSING UNITS

The previous section discussed the use of ROM and RAM IC chips. The other major types of memory chips used in PC CPUs are PROM, EPROM, EAPROM, EEPROM, and NOVRAM. Figure 3–2 shows a summary of the operational characteristics of these chips.

The PROM (Programmable Read Only Memory) chip is similar to the ROM except that it may be programmed once, and once only, by the user/programmer. In other words, the manufacturer furnishes the chip in an unprogrammed or semiprogrammed state. The user then programs the chip to his or her requirements. No erasures are possible. To change the program in a programmed PROM, you throw it away and replace it with a new, unprogrammed PROM. The PROM is seldom used because it requires special programming circuits. It does, however, have the advantage of being an unalterable backup to a ROM.

CHIP	FIXED (F) OR ALTERABLE (A)	APPLICATION	ERASABLE BY
ROM	F	Fixed Operating Memory	No
RAM	A	User Program	No
PROM	F	User Program	No
EPROM	A	User Program	UV Light
EAPROM	A	User Program	Electrical Signals
EEPROM	A	User Program	Electrical Signals
NOVRAM	A	User Program	Electrical Signals

Figure 3–2 Major Types of IC Memory Chips Used in PC CPUs

The EPROM is a PROM that can be erased. EPROM stands for Erasable Programmable Read Only Memory. The EPROM is erased by subjecting a window in its top to ultraviolet (UV) light for a few minutes.

When exposed to UV light, the chip's memory bits are reset to 0. The chip's window is covered during normal use to prevent unwanted erasure. The advantage of the EPROM is that it can be reused. The disadvantage of the EPROM is the downtime interval required for its reprogramming. Downtime includes removal time, UV light exposure time, and reinsertion time.

The EAROM is similar to the EPROM. EAROM stands for Electrically Alterable Read Only Memory. Instead of UV light exposure for erasure, an electrical signal is applied to an EAROM pin. Its advantage over the EPROM is the ease and speed with which it is reset and erased.

The EEPROM is similar to the EAPROM and is also erasable by an electrical signal to one of its pins. EEPROM stands for Electrically Erasable Programmable Read Only Memory. Technical operation is different but the action is similar to the EAPROM. Time for erasure is relatively long in computer terms. Its cycles of operation are also limited by its design. After a number of erasures, further erasing processes might not erase the entire chip. The number of effective erasures varies from chip to chip. The actual number allowable would have to be determined from the chip manufacturer's specification sheets. EEPROM is used in place of a RAM when you want fast erasure without using time for individual reprogramming of each part of the chip's memory.

The NOVRAM is a newer, combination chip. NOVRAM stands for Non-Volatile Random Access Memory. It is a combination of an EEPROM and a RAM. It is used in some advanced PCs whose operating manuals explain its operation.

MEMORY CAPACITY

Obviously, the more processes you wish to control with a PC, the more memory the PC will require. The amount of memory you will need is described in individual manufacturer's specification manuals. You will need more memory for analog control than for discrete operation of a comparable process. As memory size increases, the cost of the CPU unit will also increase. It is also possible to buy too much memory if your needs are not calculated properly. The biggest problem in determining memory size is usually planning for future process control expansion.

When matching an application to a PC, the memory required depends on the number of inputs, number of outputs, and the complexity of the control diagram. A most important feature of a PC as these factors increase is expandability of memory. Some PC models do not have memory expansion capabilities and have to be completely replaced if more memory for bigger tasks becomes necessary. However, many PC models can have memory modules added to the existing CPU. Adding a new memory module or two is much less costly than replacing the entire PC system. It is wise to consider memory expandability when purchasing a PC.

POWER SUPPLIES

The power available in most plants is 120 volts alternating current (AC) at 60 Hz (cycles per second). The PC operates on +5 and −5 volts direct current (DC). The PC CPU must therefore contain circuitry to convert the 120-volt AC input to the required 5-volt DC values. The conversion is accomplished by a built-in voltage-converting power supply. Figure 3–3 includes the make-up of a typical power supply in block diagram form. The figure also shows voltage waveforms versus time at various points in the power supply.

There are four parts shown in the diagram, plus a switching system for the battery backup system. The first block on the left, the AC conditioning block, could be included in the PC CPU. More often it is a separate external unit that is sized according to the CPU current rating. The AC conditioner "purifies" the input AC waveform. The input waveform is nor-

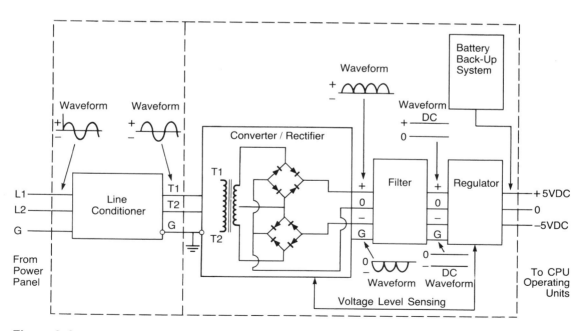

Figure 3-3
PC CPU Power Supply

mally a perfect sine wave, but can be distorted at times by two external factors. First, the power company's generating system sine wave might be distorted during system switching or by generation problems; second, equipment in your plant may cause electrical back surges that affect the purity of the electrical sine wave.

At the input from the building or power company electrical sine wave, variations can occur. The major variations are low voltage, high voltage, spikes or transients of high value, and altered waveform shape. A waveform with a spike is shown in the diagram. An AC line conditioner will eliminate, or reduce considerably, these imperfections. If passed directly to the CPU, any of these four imperfections can cause the CPU to malfunction. Only a millisecond of bad voltage waveform can cause havoc with the CPU's computer operation. The conditioner restores altered waveforms to an almost perfect sine-wave form, preventing this type of malfunction.

The second block in the diagram in figure 3–3 is the rectifier. It changes the bidirectional AC to a pulsating, unidirectional DC as shown in the waveform. Internally, a transformer steps the voltage down to an appropriate level. Then, two rectifiers produce pulsating DC outputs. One output is +5 volts and the other is −5 volts. This dual voltage is required to operate many of the IC chips in the CPU.

A computer needs a constant (not pulsating) input DC voltage for correct operation. Pulsating DC, the rectifier output waveform, falsely looks like a group of operational pulses to the CPU. Therefore, a means of smoothing it out to a reasonably constant value is required. The third block in the diagram is the filter section which accomplishes the required smoothing. The filter consists of internal circuitry, including capacitors and resistors or inductors. Alternately, the filtering may be accomplished electronically by this filtering block.

A fourth block shown in the diagram, the regulator, is usually included. A regulator keeps the voltages at or near the required 5-volt levels. The smoothed-out DC voltage received from the filter block may vary up or down from the 5-volt levels. Variations from the required 5 volts must be corrected. The regulator electronically corrects a variant voltage back to 5.0, or at least close enough to 5 volts to provide proper CPU operation.

The battery backup switch is shown on the upper right of the diagram. A switch (not shown) can transfer the output from power supply to battery. The switch is set to switch the output from power supply to battery backup power quickly and automatically if the input power supply voltage ceases. Normal power supply voltage ceases if the CPU plug is disconnected from its socket. It also ceases when building power fails. Continuity of power voltage keeps the user program from being lost, as previously discussed. Note that there is some circuitry included in the CPU to convert battery DC (for example, 22 volts) to the two 5-volt, DC-required levels. The conversion system is not shown, but could be found in the operating manual.

FIXED AND ALTERABLE MEMORY

Fixed memory is the operating system of the PC. It is contained on ROM chips which are a part of the CPU. It is set at the factory and cannot be erased or altered by the PC user.

In contrast to the fixed memory and the processor section of the CPU, the alterable memory register values are constantly being changed during PC operation. Figure 3–4 shows a typical PC CPU memory system. The individual sections, their order, and the sections' comparative lengths will vary by manufacturer and model.

The first section shown keeps track of the ladder diagram status, which is updated as the ladder diagram is scanned. The second section keeps track of input status. During each scan of the inputs, the status of each input (1 or 0) is recorded in order in this memory section. If we have 256 inputs, there must be enough memory allotted to record the 256 bits of information.

The third section keeps track of output status. After the input scan and logic diagram scan, the resulting logic output statuses are inserted in this memory section in numerical order. During the ensuing output scan,

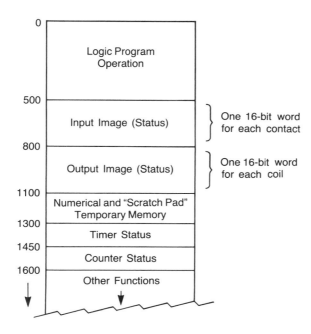

Figure 3–4
Typical PC CPU
Alterable
Memory
Make-Up

each memory output status (0 or 1) is transmitted to the output modules.

The fourth section keeps track of numerical data. It is used for temporary storage of numbers or logic in process. Temporary storage may also be considered as the scratch pad section.

Next come the individual memory sections for each function requiring variable register information. The section shown at 1300 in figure 3–4 is for the timer function. It tracks elapsed time and furnishes data to the section that analyzes time intervals by comparing them to set times. The next section is for the counter function, and the next is for number system conversion. Other functions would be continued at the bottom of the diagram as needed. The additional memory size required would depend on the total number of different functions included in the CPU.

THE PROCESSOR

The processor is the part of the PC CPU which receives, analyzes, processes, and sends information. The information, in digital pulse form, is sent and received as shown in figure 3–5. Internal logic decisions and calculations are made by the analysis section. Analysis is made in conjunction with both fixed and alterable memory. Figure 3–5 shows various operating sections for the major tasks performed.

The control section of the processor determines which operating sections are to be functional, in what order, and for how long. The actual

Figure 3–5
The PC CPU
Processor

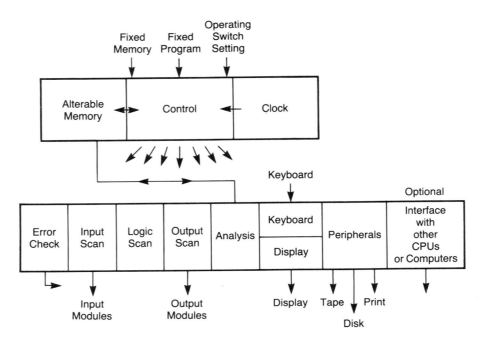

operation of this control function is quite complicated and is covered in detail in most microprocessor texts.

The input scan block, when called upon to operate, scans the inputs and places the individual input statuses in alterable memory. After analysis, the logic scan updates the output logic to the appropriate state. Next, the outputs are scanned and updated. The output statuses are changed or left alone, depending on logic analysis. Output status depends on the output status signals of the CPU. Other processor functions also take place—at a different time or at the same time. Operational intervals depend on the priority operation of the control function. In some PC CPUs, two or more computer operations can take place simultaneously. This means the PC can operate at a faster rate—which is always desirable. The time interval for operation of each section is tracked by the clock portion of the control block.

Other typical functions carried out by the processor are also shown in figure 3-5. The keyboard/display section takes action based on any keyboard operation that occurs. The display is then appropriately updated. Another section becomes active when the peripherals—tape, disk, or printers—become involved.

An important part of the processor is the error-checking section, shown on the left in figure 3-5. The error check is made on the process to detect any internal computer logic errors. Two types of error-detecting computer systems are parity and checksum. These and other error-checking systems are covered in detail in most digital logic texts.

On the far right of figure 3-5 is an optional interfacing section. This section is required if the PC is part of a larger system. This section carries out communication with other PC CPUs and a master computer, if one is used.

INPUT MODULES

The input module performs four tasks electronically. First, it senses the presence or absence of an input signal at each of its input terminals. The input signal tells what switch, sensor, or other signal is on or off in the process being controlled. Second, it converts the input signal for on, or high, to a level usable by the module's electronic circuit. For low, or off, no signal is converted, indicating off. Third, the input module carries out electronic isolation by electronically isolating the input module output from its input. Finally, its electronic circuit must produce an output to be sensed by the PC CPU. All of these functions are illustrated by the module layout in figure 3-6.

A typical input module has 4, 6, 8, or 16 terminals, plus common and safety ground terminals. The figure shows the circuit for only one terminal. All other terminals in a given module have identical circuits. The first

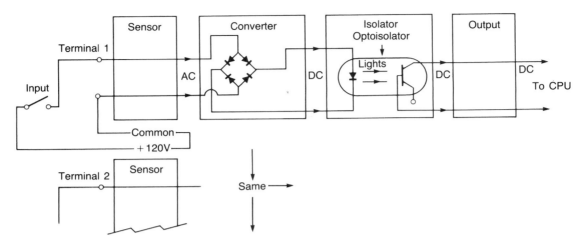

Figure 3-6
PC Input Module Layout

block, the sensor, is connected directly to the converter. The second block receives the input signal from the sensor. For AC voltage inputs, the second block, the converter, consists of a rectifier and a means to step the voltage down to a usable level. For input DC voltages, some type of DC-to-DC conversion is required within the converter block.

The output of the converter is not directly connected to the CPU. If interfaced directly, an input surge or circuit malfunction could reach the CPU. For example, if the rectifier in the converter would short out, you could have 120 volts AC fed to the CPU. Since the CPU works on only 5 volts (DC) it would be damaged. The isolation block provides protection to the CPU from this type of damage.

The isolation is usually accomplished by an optoisolator, as shown. The on-off signal is carried on a light beam in one direction. Electrical surges will not pass through the optoisolator in either direction.

The isolator, when its input is on, sends a signal to the CPU via the output block. When the isolator's output is on, it is sensed by a coded signal from the CPU. Each module is assigned a coded series of numbers by its DIP switch settings (for example, 9 through 16), as discussed in chapter 2. Each terminal number of the module is assigned a number in consecutive order. The on-off status for each number is checked on each sweep of the input scan. The result, on or off, is placed in the alterable memory, as previously discussed.

OUTPUT MODULES

The output module operates in the opposite manner from the input module. A DC signal from the CPU is converted through each module section (terminal) to a usable output voltage, either AC or DC. A block diagram of the output module is shown in figure 3-7.

A signal from the CPU is received by the output module sensor, once for each scan. If the CPU signal code matches the assigned number of the module, the module section is turned on. The identification numbers of the module are again determined by the setting of the module DIP switches, as discussed in chapter 2. As with input modules, there are 4, 6, 8, or 16 terminals or sections. If no matching signal is received by a terminal during the output scan, the module terminal is not energized.

The matching CPU signal, if received, goes through an isolation stage. Isolation, again, is necessary so that any erratic voltage surge from the output device does not get back into the CPU and cause damage. The isolator output is then transmitted to switching circuitry or an output relay. AC switching is usually accomplished by turning on a triac. The output of a module section may be through a relay, or a DC or AC output. All three types are shown in figure 3-7.

All terminals of a single module have the same output system. In other words, an 8-terminal module would not have some AC and some DC outputs or voltages of differing values. All would be the same.

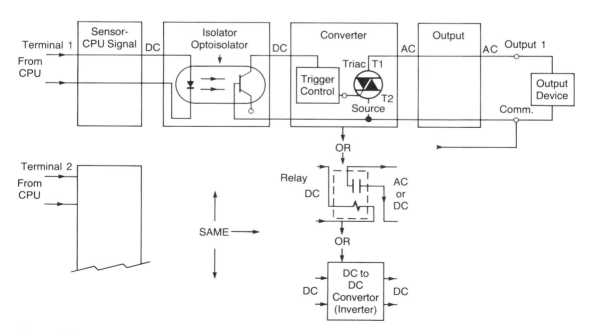

Figure 3-7
PC Output Module Layout

EXERCISES

1. Describe how the following sequence is carried out by a PC CPU:
   a. A program on a disk is placed in the program memory. The program recalled from the disk includes a line where switch 34 causes output 54 to go on.
   b. Switch 34, which is connected to input 34, is turned on.
   c. The CPU recognizes that the switch is on.
   d. The CPU logic turns the internal output 54 on.
   e. The 54 output status is conveyed through the output module, terminal 54, to an external light.
   f. Next, two alterations are made in the ladder program.
   g. The entire logic program is then recorded on a printer.
2. Obtain one or two manufacturer's manuals. Determine the following from the manual or manuals:
   a. Size of memory
   b. Memory map. Which sections of memory are used for what?
   c. Input module system and output module system of operation.
3. List five types of IC chips commonly used in a PC CPU. Describe the operation of each.
4. Describe the feasibility of the following IC chips: EROM, ERAM, EPROM.

Installation and Testing

4

At the end of this chapter, you will be able to
☐ List and discuss the procedure for checking the parts of a PC as received from the manufacturer.
☐ Describe the procedure for assembling and interconnecting the PC system. This includes the setting of various switches in the PC system, including the I/O switches.
☐ List environmental factors which may have an effect on PC operation.
☐ List the reasons for PC grounding and suppression and how they both are accomplished.
☐ Describe a complete testing procedure for a newly received PC.

INTRODUCTION

This chapter discusses the installation and testing by the user of a new PC. Proper installation and thorough testing before the PC goes on-line ensures its dependable and continuous functioning. Included in the discussion is a check on the condition of the PC upon receipt from the manufacturer and the consideration of the environment in which the PC is to operate. Electrical installation is covered, including grounding and suppression requirements. A master safety shutdown circuit will also be discussed, as will proper testing procedures.

RECEIVING CHECK

When you receive the PC system from the manufacturer, inspect the packing boxes for any obvious damage. If the boxes are damaged, take a picture of them before opening, in case the parts inside the packages are also damaged. Then, inventory the parts and manuals received against the packing list provided. Also, review the purchase order. Its listing of parts ordered may differ from the parts received. Record any discrepancies between the packing list, purchase order, and parts received. If equipment is damaged, broken, or missing, the supplier should be notified. In industrial organizations, the notification is made through the company purchasing department. A disposition to rework, replace, or return must be mutually agreed upon.

CONSIDERATION OF THE OPERATING ENVIRONMENT

The factors in this section should be considered to ensure continuous, reliable operation of the PC system after installation:

Enclosure. The PC can be installed in the open; more often, however, it is installed in an enclosed, NEMA-type metal enclosure. NEMA, the National Electrical Manufacturers Association, sets standards for the sizes of enclosures to meet installation codes. The NEMA enclosure must be planned to allow adequate room for the incoming control wires and power wiring and so that all parts and wires are easily accessible for installation, future alterations, and troubleshooting. Appropriate racks are needed to support groups of wires throughout the enclosure. The enclosure should be large enough to allow for future expansion.

Temperature. The PC has upper and lower temperature operating limits, normally 0 °C (32 °F) and 60 °C (140 °F). These limits must not be exceeded during plant operation or during seasonal temperature changes affecting the PC's ambient temperature. For example, a PC installed over an annealing oven can soon develop operational glitches due to excessive heat, especially during the summer.

Moisture, dust, and corrosive atmosphere. A PC may be required to operate in an area of high humidity. Consideration must be given to the level of moisture, which, if too high, can cause electrical and electronic malfunctions. Dust can clog cooling ports and create paths for electrical shorts. In corrosive atmospheres, which could occur in such operations as chemical plants or refineries where oxidizing fumes may be present, electrical connection points can fail due to the build-up of oxides on the wires and terminals. In all three cases, suitable protected enclosures must be used, as specified in the National Electrical Code according to corrosion type.

Vibration. If the CPU is subjected to excessive vibration, transmitted from nearby vibrating equipment, it can malfunction. Vibration can also cause early CPU failures and reduce the life of the PC equipment. Vibration effects can be reduced by shock-prevention mountings.

ASSEMBLY

Many electronic parts and assemblies are easily damaged by small charges of static electricity. To protect these parts from such damaging static discharges, the manufacturer will normally ship them in anti-static bags. When removed from the bag, these parts require special handling. Units and modules received in these bags must first be inspected for damage. If the bags are damaged, return them to the supplier for replacement or for recheck. The parts should then be removed from the bag in a static-free environment. When the modules are installed into the system, the same precautions are necessary. Portable grounding kits are available from some PC manufacturers to prevent static damage of parts during handling.

Practically all PCs, even lower priced ones, have backup battery systems. Some battery systems use a common 1.5- or 9-volt, long-life battery. Others use various types of batteries with special voltage ratings. Some PCs use a rechargeable battery which is trickle-charged by a small power supply in the CPU. Not all batteries are connected when shipped; they may be in place but insulated from the battery clips by two spacers that are removed for PC operation. In other cases, separately shipped batteries are installed according to the manufacturer's instructions. Special precautions during installation might include removal of some modules or wires to prevent static damage or electrical surges to some part of the CPU. In all cases, battery voltage should be checked for compliance to voltage specifications listed in the PC manual before installation.

All PC systems have at least one fuse; many have a number of different fuses. These may be in place when the PC is shipped. If not, the fuses must be installed according to start-up instructions in the manual.

ELECTRICAL CONNECTING, GROUNDING, AND SUPPRESSION

Once you have installed all the parts, you can plug in the line cord for the CPU. Check the CPU for proper operation as you turn the keyswitch or master switch from position to position. Check to see that all operating pilot lights come on at the proper time.

If the PC CPU does not operate properly at this point, internal visual checks are in order. The faceplate or an appropriate panel may be removed. You can then make a visual check for any loose connections.

Assembly of the input and output modules is now required on larger units. The individual modules are placed on the racks furnished by the manufacturer. The racks are not only for mechanical support, but have interconnecting electrical wires and connections in them, as well. Take care that modules are put precisely in place.

The input and output modules are next connected to the CPU with the proper cables. Care must be taken that the connecting ribbon cables are not too twisted or pulled during installation. The power should be off for this procedure. Module switches are then set as described in chapter 2. Next, the wires from the external devices and switches are attached to the I/O terminals. The incoming wires from the input and output devices

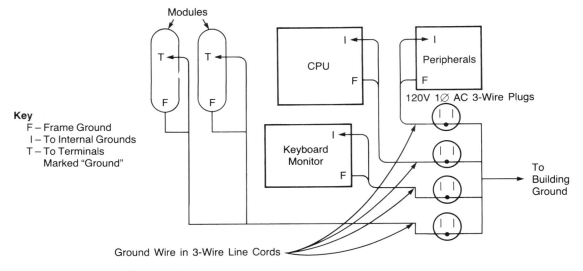

Figure 4-1
PC System Grounding Scheme

must be fastened to the terminals securely. The standard practice of "hand tight" is normally followed.

Peripheral devices such as printers, disk drives, and tape drives may now be interconnected to the system by means of their cables. Remote stations and busses to other PCs and computers should not be connected until the individual PC check-out is completed.

Proper electrical grounding of the wiring of the equipment and cabinets is essential for personnel safety and to assure proper equipment operation. An ungrounded or improperly grounded wire or part could become shorted electrically to a metal cabinet or rack, presenting an electrical shock hazard to users. In addition, the PC is computer-based, and computers need a proper and solid grounding system for consistently trouble-free operation. A typical wiring scheme for grounding is shown in figure 4-1.

Electrical disturbances from devices outside the PC system can cause program operation malfunctions. Solenoids, starter coils, motors, and certain other devices are electrically inductive in nature. When these inductive devices are energized or deenergized, they can cause an electrical pulse to be back fed into the PC system. The back-fed pulse, when entering the PC system, can be mistaken by the PC for a computer pulse. It takes only one false pulse to create a malfunction of the orderly flow of PC operational sequences.

Electrical disturbances in the air also can create false pulses to the PC. These disturbances can be reduced or eliminated by the use of shielded interconnecting cables. A stranded copper outer sheath around the shielded cable prevents the disturbance pulse from getting to the cable wires inside. The disturbance can also enter through the wires themselves. These direct disturbances can be reduced or eliminated through suppression techniques. Some of these reduction methods are shown in figure 4–2. Essentially, the suppressor absorbs the inductive-caused electrical disturbances. Therefore, no disturbance signal remains that can be sent back into the PC.

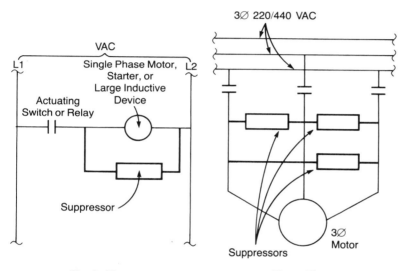

Figure 4–2
Input and Output Suppression Techniques

Single Phase   **Three Phase**

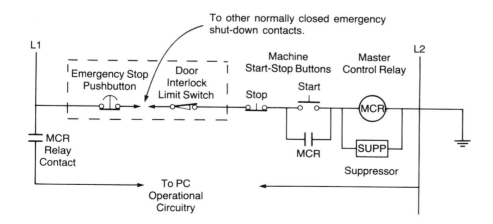

Figure 4–3
Master Control Safety Shutdown Scheme

Many systems have a master control relay system for safety shutdown of the PC operation. This is an override of the whole PC program. When on, the safety shutdown allows the PC to operate. When the override is deenergized the PC will not operate. A typical master shutdown system is shown in figure 4-3. If control power fails, the PC operation is shut down.

TESTING

When completely assembled, the PC is ready for testing. Testing may be accomplished in any one of three modes. First, the PC can be tested "as is," without attaching any wiring to the I/O modules. Second, it may be tested with a simulator (illustrated in appendix B). Third, it may be tested after it is hooked up to the system it is to operate.

In any testing procedure, documentation of all tests and their results should be recorded in writing for later reference. More important, a review of the documentation will assure that all necessary tests have been run.

When the PC is tested with no wiring attached to it, electrical jumpers must be used to energize the inputs. A jumper is moved around from input to input to check for correct operation. Output operations, which then are program energized, are indicated by the operation of the corresponding indicating lights on the output module. Alternately, the FORCE mode (see chapter 14) may be used to check for correct input operation. Instead of moving jumper wires around for the actuation of inputs, the keyboard is utilized. The disadvantage in using FORCE as a simulation is that input module operation is not checked, since the inputs are only simulated. The FORCE mode will be described in detail shortly.

All testing should be performed in the MONITOR mode (see chapter 14). In this mode you will be able to observe the ladder program operation on the PC screen, which gives a better view of the PC internal operation.

The second test method, which uses a simulator, is performed similarly to the first method; the jumper wires, however, are not necessary. A switch is attached for each input, and indicating lights are attached for each output.

The third method of PC testing is to connect the PC with the factory operational system connected to the PC. This method has one major disadvantage, however. If the PC equipment is malfunctioning, the equipment being used for testing can be damaged. Furthermore, an operator or programmer error can cause sequence problems or even damage to equipment. Personnel in the area under the PC's control could be injured during any malfunction.

The FORCE mode (discussed in detail in chapter 14) is often used to test the PC for proper operations for any of these three test methods. The

FORCE mode requires turning inputs and outputs on or off from the keyboard. This overrides the system's normal operation through the input module. Therefore, the use of the FORCE mode could be dangerous to equipment or personnel. Some piece of equipment or component could be turned on unintentionally through the keyboard.

So far, the PCs we have discussed have discrete on-off input and output modules and systems. Some PCs have analog capabilities, which means that inputs and outputs have continuously varying values. Testing these analog PCs requires special equipment and procedures. A special simulator is mandatory for testing a system with other than discrete modules. For these nondiscrete analog modules, the on-off light indicators are of no help in determining proper CPU and output operation. An on-off light does not indicate a variable value. Analog simulators are available as illustrated in appendix B.

Testing of peripherals such as printers, disk drives, and tape drives also takes special testing procedures. Each peripheral device should be completely tested in all possible operating modes. For example, if a printer can print five different types of information, all five modes of operation should be tested.

A complete test of the PC system and CPU involves checking every function, not just the ones to be used on the immediate process application. Each input and output should be checked—all 200 if there are 200. Also, every function (for example, Timer, Counter, Master Control Relay) should be operated and observed. Check each function, even if it is not to be used initially. Once the PC warranty runs out, it is probably too late to have a malfunction fixed at no cost. Furthermore, an untested function that is later found to be faulty could cause delays until it is fixed.

EXERCISES

1. Choose a local industry process where a PC might be installed. List and describe what environmental factors must be considered before installing the PC system for control of this process.
2. Obtain a manual from a PC manufacturer that includes an installation and testing section. List in order the operational steps to install the equipment properly.
3. From the manual you used in exercise 2, draw a block diagram showing how grounding is accomplished for all PC system components.
4. Do electrical resistance heaters controlled by a PC require electrical suppression? Why or why not?
5. From the manual you used for exercise 2, explain the PC's safety shutdown or MCR system.

6. From information from the manual you used for exercise 2, list in order the necessary steps for completely testing the PC after installation. Extra credit: Make up a check list for carrying out the testing and for recording the results.

Troubleshooting and Maintenance

5

At the end of this chapter, you will be able to

☐ List and describe troubleshooting procedures for general electromechanical devices.
☐ List and describe specific PC troubleshooting procedures.
☐ Describe corrective action and documentation for common PC failures.
☐ List and describe general and preventative maintenance procedures for PCs.

INTRODUCTION

There are universal troubleshooting procedures that apply to all electromechanical devices, PCs included. There are also numerous unique troubleshooting procedures that apply to PCs specifically. Both general and specific troubleshooting procedures will be discussed in this chapter. Troubleshooting charts and procedures typical of PC manuals will be listed and discussed, along with some procedures for general and preventative PC maintenance.

GENERAL TROUBLESHOOTING PROCEDURES

Some troubleshooting procedures have general applications to all electromechanical devices, and PCs are no exception.

Safety is the primary consideration during troubleshooting procedures. Will the process under control by the PC come on unintentionally during the equipment checkout? If so, the process must be operationally locked out in some manner. Turn off the power when working inside of the device unless power is necessary for the analysis. Wear safety glasses during circuit checking in case of an electrical flash or small electrical explosion. Appropriate insulated electrical tools should be used. Most important, work deliberately and consider the consequences of each step of the checking procedure.

Once general safety precautions have been taken, you are ready to proceed into malfunction analysis. One major step is to understand exactly what is malfunctioning. Try to get more than one opinion on the problem, if possible. Document the problem in detail, including the date, time, severity, and circumstances of the malfunction, and include a second opinion if one can be obtained. Then, verify that the reported problem actually exists before trying to solve it. Once the malfunction is determined, begin checking. Check for and replace any blown fuses. Next, a complete visual inspection may reveal the cause of the malfunction. Is there power to all control circuits? Are there any broken wires? Are switches set in the proper positions? Have any undocumented alterations to parts been made? A review of terminal wiring diagrams may reveal that an unrecorded change is causing the trouble.

SPECIFIC PC TROUBLESHOOTING PROCEDURES

As stated earlier, some troubleshooting procedures are exclusive to PCs. This section describes some of these unique procedures.

First, establish that the system malfunction is not caused by an external part or system. Then, check the PC itself for proper operation in each applicable mode. Larger PCs have an available screen readout for CPU status. Call up the status and refer to the operating manual to determine what the CPU display should look like to indicate proper operational conditions. Any portion not meeting the norm shows an operational problem area. In some cases, just clearing the CPU memory and reprogramming the PC will eliminate the malfunction.

One way to analyze a PC problem is by sub-unit substitution. If you have another CPU available, replace the one in service with it. Reprogram the newly inserted CPU with the program the old CPU was using. If the problem is corrected, the removed CPU could be the culprit. If you determine that the original CPU is not the problem, analyze other PC subparts by substitution in a similar manner.

Use of the MONITOR mode is helpful throughout the troubleshooting analysis. By observing the program ladder operation on the screen, any mis-operation may be discovered. As in testing, the FORCE mode is useful in simulating operating conditions. A word of caution, however: a portion of the program may operate properly in the FORCE mode but not during actual operation. For example, input IN 0045 may operate correctly in the FORCE mode, but not in actual operation. This would indicate that the input, IN 0045, is malfunctioning—for internal or external reasons.

In addition to the CPU status screen display, many PCs have available a fault indication register display. The fault display may appear automatically or it may have to be called up. A typical fault display is shown in figure 5–1. The figure includes a display and an interpretation sheet, which is found in the operating manual. Other displays with more

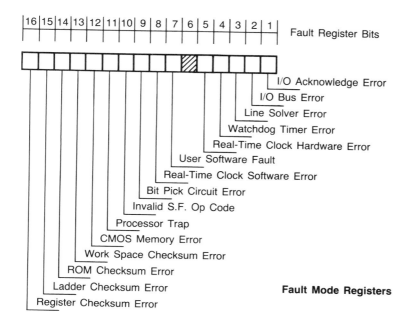

| 16 | 15 | 14 | 13 | 12 | 11 | 10 | 9 | 8 | 7 | 6 | 5 | 4 | 3 | 2 | 1 | Fault Register Bits |

I/O Acknowledge Error
I/O Bus Error
Line Solver Error
Watchdog Timer Error
Real-Time Clock Hardware Error
User Software Fault
Real-Time Clock Software Error
Bit Pick Circuit Error
Invalid S.F. Op Code
Processor Trap
CMOS Memory Error
Work Space Checksum Error
ROM Checksum Error
Ladder Checksum Error
Register Checksum Error

Fault Mode Registers

Figure 5-1
CPU Fault
Mode Register
Display and In-
terpretation
(Courtesy of
Westinghouse)

Bit	Indicated Fault	Suggested Action
1 to 3	*See Table 9-1.	*See Table 9-1.
6	1 = I/O Image Memory Error	**See Flow Chart #7.
7	1 = Real-Time Clock Error	**See Flow Chart #8.
8	1 = Line Solver Error	
10	1 = Watchdog Timer Error	This fault normally results from: 1. A program that takes longer to execute than the 100 msec the processor allows. 2. Possible cause: a program with too many complex functions being performed on the same scan. If this fault occurs during programming installation and checkout, re-examine the program and reprogram, as necessary. If this fault occurs after a program has run successfully for an extended period of time, the program may not be at fault. In this case, perform the troubleshooting procedure in figure 5-2.

Fault Register
Interpretation

*Table not included in this text.

**Typical flow charts shown in Figure 5-2.

specific information are available on some PCs. A message on one of these PCs might say "OUTPUT 0024 IS SHORTED" or "REGISTER 043 IS NOT WORKING," for example. More sophisticated PC systems have messages that also tell which external devices are not working and why (for example, "MOTOR NUMBER 45 IS OVERHEATING").

Another PC troubleshooting aid is sequential listings of troubleshooting steps in an operating manual. The lists can be in consecutive form, as shown in this example:

1. Turn on operating switch number 7
2. Key switch to run
3. Light number 4 must light
4. If light number 4 does not light, check bulb
5. If bulb is OK, put all 1's in Address 909, etc.

Other sequential listings are in computer flow diagram form. See figure 5-2 for an example. By following the flow diagram, the subsystem may be checked for correct operation or determination of a malfunction.

CORRECTIVE ACTION

Upon determining the cause of the PC malfunction, corrective action is taken. Usually, corrective action involves replacing a faulty part with a new part. The faulty part is usually a printed circuit board or a single electronic component. Boards are usually replaced in their entirety in these cases. A PC printed circuit board must be tested extensively to analyze failures caused by the board's faulty components; these tests are possible only in the manufacturer's facilities.

If the system still does not work when a faulty part is replaced, there are three possible reasons. First, the part used for replacement may also be faulty. A replacement part recheck is required. Second, there may be another part or parts in the unit which are also faulty. Further analysis is needed. Third, the replacement part fails when it is installed because the original defective part had failed due to an overload caused by some other malfunction of the system. The replacement part has been caused to fail by the same problem. If you have replaced a fuse twice there isn't much loss; if you have replaced a $1500 circuit board, a second failure becomes quite expensive.

For expensive parts, some type of pretest is in order before replacing a failed part; for example, a quick precheck before a circuit board replacement to determine if its supply voltage is correct.

A written log of failures and corrections for each PC should be kept. The log will show if there is any failure pattern after a number of entries are made. Any pattern of failures will show which PC problems may be

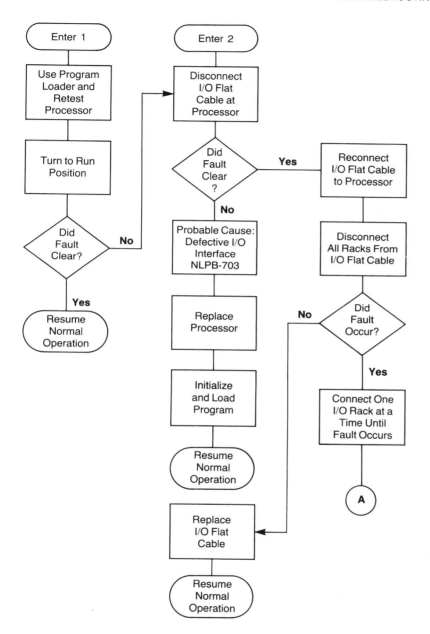

Figure 5-2
PC Trouble-
shooting Flow
Chart (Courtesy
of Westing-
house)

anticipated. More spare parts for a particular failure mode may have to be kept on hand. Another reason for the log is historical in nature. If a failure occurs a second time, reviewing the log will enable the trouble-shooter to benefit from past experience. The log should include a description of the failure and the corrective action taken, as well as the date and shift (first, second or third) of occurrence.

PC MAINTENANCE

Maintenance and preventative maintenance of a PC include:

1. Periodic checks on the tightness of I/O module terminal screws. They can become loose over a period of time.
2. Check periodically for corrosion of connecting terminals. Moisture and corrosive atmospheres can cause poor electrical connections. Internally, printed circuit board end connectors may become corroded also.
3. Make sure that components are free of dust. Proper cooling of the PC is impossible through a layer of dust.
4. Stock commonly needed spare parts. Input and output modules are the PC components that fail most often. Stocking is especially essential if there is no convenient manufacturer's service station and parts depot. Proper levels of spare parts inventory is a trade-off between costly inventory and prolonged downtime without parts.
5. Keep a duplicate record of operating programs being used. These may be recorded on paper, tape, or computer disk. These records should be kept in a plant location away from the PC operational area. Copies of long, expensive programs should be kept off the premises to prevent their loss in case of fire or theft.

EXERCISES

1. List and describe the major safety rules for PC troubleshooting.
2. List and describe some general troubleshooting rules for electromechanical devices.
3. List and describe specific troubleshooting procedures with specific application to PCs.
4. Set up an effective log system to record PC failures and corrections.
5. List and describe the major areas for PC maintenance and preventative maintenance.

NOTE: All chapter exercises can be answered more fully by reviewing the applicable sections of operating manuals for PCs from one or two different manufacturers.

SECTION TWO

PROGRAMS AND SOFTWARE

Creating a Ladder Diagram for a Process Problem

6

At the end of this chapter, you will be able to

☐ List the major steps in creating a PC program for an industrial situation.
☐ Describe the content of each of these steps.

INTRODUCTION

Planning without action is a waste of time and money, and action without planning creates chaos. The purpose of this chapter is to outline some of the planning needed to create good, workable, safe PC programs—without chaos.

You may want to omit this chapter for now if you work with preprogrammed PC programs. You should include this chapter if you have to create your own programs, modify programs, or if you doubt the validity of the program you have.

Chapter 6 is written in relay logic. The principles are readily converted to PC programs, as will be covered in chapter 11 for coils and contacts, and in chapter 12 for timers.

CIRCUIT CONSTRUCTION PLANNING STEPS

Some of the steps in planning a program are:

1. Define the process to be controlled.
2. Make a sketch of the process operation.
3. Create a written step sequence listing for the process.
4. Add sensors on the sketch as needed to carry out the control sequence.

5. Add manual controls as needed for process setup or operational checking.

6. Consider the safety of the operating personnel and make additions and adjustments as needed.

7. Add master stop switches as required for safe shutdown.

8. Create the ladder logic diagram that will be used as a basis for the PC program.

9. Consider the "what if's" where the process sequence may go astray.

Some other steps needed in program planning that we will not cover are troubleshooting of process malfunctions, parts list of sensors, relays, etc., and wiring diagrams, including terminals, conduit runs, etc.

CHAPTER EXAMPLE—THE NINE STEPS

To illustrate the nine steps of the planning sequence, we shall use a fundamental industrial control problem. We will then go through the creative process to illustrate each of the steps of the planning process.

Step 1

Define the problem.

We wish to set up a system for spray-painting parts. A part is to be placed on a mandrel. When the part is in place, the mandrel automatically raises the part into a hood. After the part rises and is in the hood, it is to have spray paint applied for a period of six seconds. At the end of the six seconds, the mandrel returns to the original position. The painted part is then removed from the mandrel by hand. (We shall assume for our illustration that the part dries very quickly.)

Step 2

Make a sketch of the process. (figure 6–1).

Step 3

List the sequence of operational steps in as much detail as possible. The sequence steps should be double- or triple-spaced so that any omitted steps discovered later may be added. The following is a step sequence for this process:

1. Turn on paint pump and pneumatic air supply.

2. Turn system on. Requires extra push buttons separate from system buttons.

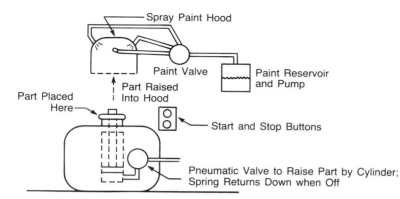

Figure 6-1
Sketch of the
Spray Process
System

3. Put part on mandrel. Sensor indicates part in place.

4. Push system start button or buttons. Having to push two buttons with both hands reduces the possibility of the operator's hands being injured by the rising mandrel.

5. Mandrel is raised by a pneumatic cylinder energized by the opening of an electrically actuated air valve. The mandrel will return down by gravity and downward spring action when the valve is reopened. Note that when the mandrel rises, the part in the place sensor at the bottom becomes deenergized.

6. When the part reaches the top, it is held against a stop by the air pressure. A sensor indicates the part has reached the top.

7. A timer starts and runs for six seconds.

8. During the timing period of six seconds, paint is applied by the sprayer.

9. At the end of the six seconds, painting stops and the part lowers.

10. Up sensor is deenergized when the part leaves the top.

11. Part arrives at the bottom, reenergizing the part in the place sensor. (We assumed that the part in the place sensor did not rise with the mandrel.)

12. Part is removed from the mandrel.

13. System is reset so that we may start at step 3 again.

Step 4

Add sensors as required. Once we list the sequence, we find that sensors are needed in the machine to indicate process status. We need a sensor (LSP) to show that the part has been placed on the mandrel initially. We also need a sensor (LSU) to indicate when the mandrel is fully extended upward. Among other possible sensors that a process such as this might need is one to make sure the paint sprayer has paint and one to make sure the inserter's hand is out of the way. Depending on the process and the detail of control, there could be other sensors required, as well. Figure 6–2 includes the two basic sensors, LSP and LSU, and their locations. The figure also shows the enclosures needed, along with the locations of start and stop buttons.

Figure 6–2
Sensor, En-
closures, and
Push Button
Locations

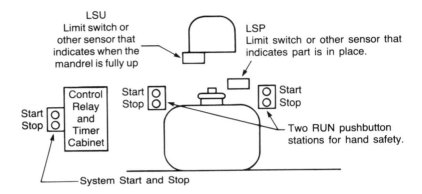

Step 5

Add manual controls as needed. Perhaps we would require a manual push button to raise the mandrel to the top for setup purposes. The manual up position is needed when we set the spray-gun pressure for optimum paint coverage. We will include push button up (PBU) on our ladder diagram to accomplish this manual control.

Step 6

Consider the safety of the machine operator. One basic way to keep hands out of a process is to have two start buttons. Then both hands must be away from the work to depress both buttons (which works until the operator figures out how to use one knee and one hand). Other considerations, which we will not cover in detail, might be operation of a fan to disperse fumes during spraying, or perhaps a photocell proximity-personnel-system-stop device.

Step 7

Add master stop switches as needed for operator safety. This may seem to be part of step six since both steps deal with operator safety. It is a continuation of the safety issue, but emergency stop switches are so important that they need special consideration as an additional step.

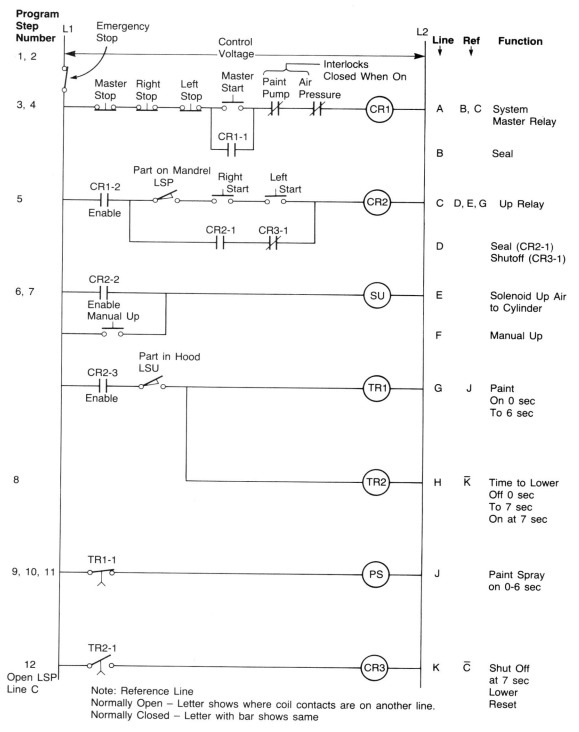

Figure 6-3
Ladder Logic Diagram

Step 8

Create the ladder logic diagram. The diagram created is to include the steps and considerations of the first seven steps. This is shown in figure 6-3 for our spraying example.

Step 9

Determine the "what if's," or potential problem areas. After the ladder diagram is completed, all possible situations and emergencies should be listed. In this example, some of them might be:

☐ What if no part is in place when the start buttons are pushed?

☐ What if the power fails during the cycle, when the part is rising, during painting, or at any other time?

☐ What if the sprayer runs out of paint?

☐ What if the same part is left in for a double coat?

☐ What if the stop button is pushed? Does the stop button really stop the entire process, or can the mandrel move and create a safety problem after the stop button is depressed (it can).

All of these types of questions should be considered in the final sequence and ladder diagram. Review of our ladder diagram in figure 6-3 covers some of these contingent situations, but not all of them. Further modifications would be needed for a more complete consideration of contingencies.

EXERCISES

Solve the following problems using the nine-step planning sequence:

1. A part is placed on a conveyor. The part automatically moves down the conveyor. In the middle of the conveyor the part goes through a two-foot-long painting section. The sprayer paints for the time the part is under the booth, during which time the conveyor does not stop. When the part reaches the end of the conveyor, the conveyor stops and the part is removed. Assume that only one part can be on the conveyor at one time. (HINT: use one limit switch at the front of the booth and another at the end.)

2. Same as exercise 1 except that the part stops in the middle of the conveyor and is stamped, not painted, and then continues to the end of the conveyor.

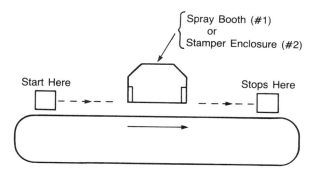

Figure 6-4
Diagram for Exercises 1 and 2

General Programming Procedures

7

At the end of this chapter, you will be able to
- [ ] Describe a typical PC keyboard layout and its operational procedures.
- [ ] Describe the difference between legal (proper) and illegal (improper) PC ladder programming layouts.
- [ ] List the important considerations of program scanning rate and sequence, and their effects on system operation.
- [ ] Describe what action to take when a PC operational fault occurs.
- [ ] State why some PC circuits can be operationally unsafe and describe how a corrective "fail-safe" circuit can be added to an operating system.

INTRODUCTION

This chapter explains the use and programming of the PC equipment discussed in chapter 2. Chapter 2 described various hardware parts of the PC system; chapter 7 will detail the use of some of the PC equipment. The descriptions apply to PCs in general; reference to the manufacturer's manual is required in all cases for complete, specific understanding of a given PC's proper operation.

Chapter 7 will cover keyboards, ladder diagrams, program formatting, and proper layout of the control diagrams. Diagram scanning sequence considerations, fault light interpretation and action, and operational safety of the system will also be discussed.

TYPICAL KEYBOARD LAYOUTS

Chapter 2 illustrated some typical keyboard layouts ranging from large to small. In small hand-held units, the keyboard is small and the keys have multiple functions. Multiple-function keys work like second-function keys on calculators.

In larger monitor systems, a "menu" system is often used instead of dual function keys. Figure 7–1 shows the detail of a large keyboard.

When certain keys are depressed, a number of blocks with words appear on the bottom of the screen. Then, a second key directly below the

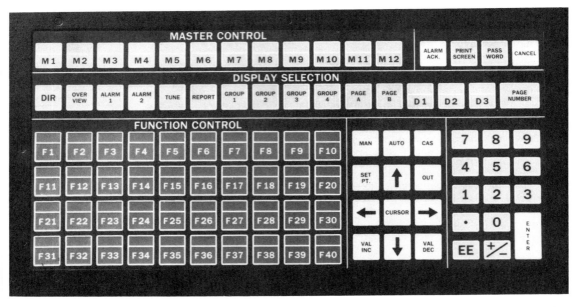

Figure 7-1
Large PC Keyboard (Courtesy of Texas Instruments)

Figure 7-2
PC Monitor
Screen with a
Menu Appearing
(Courtesy of
Giddings and
Lewis)

```
** MODE MENU **

1. PROCESS SCREEN

2. HEAT CONTROL SETUP

3. MOLD AXIS SETUP

4. SCREW SETUP

5. PRESS FAULTS

6. ROBOT FAULTS/DIAGONISTICS

7. ROBOT REAL TIME DISPLAY

CL. RETURN TO MAIN MENU


** PRESS THE DESIRED MODE KEY **
```

blocks must be chosen and depressed. See figure 7–2 for a screen with a typical menu appearing on the CRT. Individual keyboards and their operational systems vary among manufacturers. The individual manuals describe their operation.

PROGRAMMING FORMATS

Certain chapters throughout the book will show some different manufacturers' format approaches to controlling processes. We will use a general format similar to those of companies having a major share of the PC market at present. Experience has shown that when an individual learns to program one type of PC, that individual can easily master other PC systems, even though the formats are somewhat different.

Some of the factors that vary between formats are generally nomenclature, numbering schemes, and screen appearance. (Nomenclature descriptions will be covered in examples in individual chapters.) Another format variation is in the numbering formats for contacts, outputs, and registers. These formats include letters, numbers, or a combination of both. Individual PC operating manuals explain the various systems of designating functions and registers.

The major variation in PC monitor screen appearance is between hand-held types and full-monitor types. Hand-held or small keyboard-monitor systems show only a part of the ladder rung at a time. Larger systems show the entire ladder rung or rungs on the screen. This larger screen gives a complete picture to the viewer, but at a higher cost. For training purposes, in most cases a larger monitor is preferred, because being able to view the whole picture at one time aids the learning process. After learning PC programming on a large monitor, it is fairly easy to learn how to program a smaller, hand-held unit. As a trade off between cost and convenience, screen size is a factor to consider before buying a PC.

A typical keyboard sequence for one output depending on two inputs being switched on might be:

1. Turn PC on.
2. Clear PC memory.
3. Choose EDIT mode.
4. Press "contact" (normally open).
5. Assign number by pressing numerical keys.
6. Press return or enter, depending on PC model.
7. Repeat 4, 5, and 6 for second input contact.
8. Continue line to the right.
9. Press "coil."
10. Assign number by pressing numerical keys.

11. Press return or enter, depending on PC model.
12. Enter ladder into memory by pressing insert, or equivalent.
13. Now, turning on both inputs should turn the output on. Output is on only if both inputs are energized (contacts closed).

PROPER CONSTRUCTION OF PC LADDER DIAGRAMS

A PC programming format's limitations must be observed when programming a PC ladder diagram. Otherwise, the PC CPU will not accept the screen-programmed ladder diagram into its memory. In some cases of incorrectly formatted ladder diagrams not being received, an error message will appear on the screen. The error message will show that the program

Figure 7-3
Proper Diagram
Nesting Required
Orientation

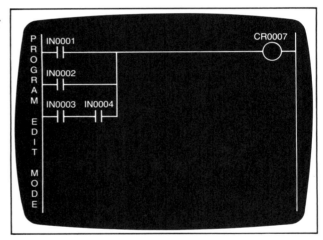

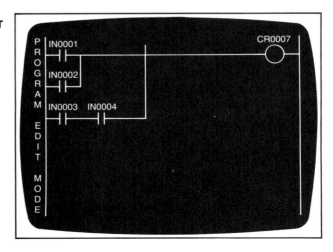

was not entered, and why. Why might the ladder diagrams be incorrect for a PC? Some of the ladder construction limitations for typical PCs are:

1. A coil or output must be inserted first.
2. A coil or contact must be inserted last.
3. Only one output and its contacts appear on the screen at a time. In others, up to five may appear, depending on screen limitations.
4. Only one output may be connected to a group of contacts.
5. A contact must always be inserted in slot 1 in the upper left.
6. All contacts must run vertically. No horizontally oriented contacts are allowed.
7. The number of contacts per output is limited. For example, 8 across by 10 down, or 11 across by 7 down, etc.
8. Contacts must be "nested" properly (see figure 7-3). The "nesting" requirements will vary from one PC to another. The figure shows one manufacturer's required format.
9. Flow must be from left to right (see figure 7-4).
10. Contact progression should be straight across (see figure 7-5).

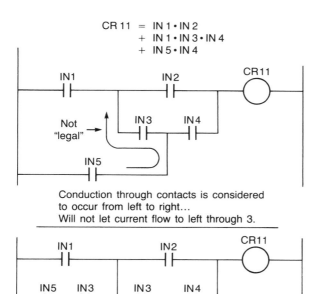

$$CR\,11 = IN\,1 \cdot IN\,2$$
$$+\ IN\,1 \cdot IN\,3 \cdot IN\,4$$
$$+\ IN\,5 \cdot IN\,4$$

Conduction through contacts is considered to occur from left to right...
Will not let current flow to left through 3.

The addition of 2 contacts (IN5 and IN3) *adds* the path IN5, IN3, IN2.

Figure 7-4
Proper Diagram
Flow Orientation

Figure 7–5
Proper "Straight
Across" Orien-
tation for Contact
Insertion

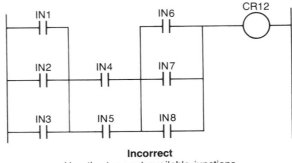

Incorrect
Use the topmost available junctions

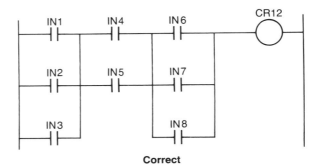

Correct

Again, the individual operational manuals contain information on the proper programming of a given PC system.

PROCESS SCANNING CONSIDERATIONS

All PCs take a discrete amount of time to scan their operational programs completely. The scanning takes place left to right across each rung and from the top to bottom rungs, in order. Typically, the complete ladder scan time is a few milliseconds. Early computers took as long as a few seconds to make a complete scan. While the present-day, microprocessor-based PC scans much faster, its speed must often be considered. For example, we might have a critical safety point in the diagram which must be monitored twice per millisecond. Suppose that scan time is five milliseconds. The critical safety point is therefore only checked out once every five milliseconds, not the required once every half millisecond. There are advanced techniques to handle this programming problem. The technique usually used is the UPDATE IMMEDIATE function, which will be described in chapter 24.

Another scanning consideration involves proper operational sequencing of events. An output might not go on immediately in a sequence as it would in a relay logic system. In a relay logic system, an event occurring anywhere in the ladder control system results in immediate action. In a PC ladder control diagram, however, no effect takes place until the rung is scanned. In most cases, this PC logic delay effect is inconsequential. However, in fast-acting, interlocked, or rapidly sequenced PC programs, the elapsed time required for scanning must be considered. For example, in figure 7–6 we see that the on-off status of output is identified as CR 0062. CR 0062 is controlled by two contacts, CR 0053 and assigned input number CR 0317. The input switch connected to IN 0015 is closed. CR 0053 on line A is turned on by the contact IN 0015. Then, the CR 0053 contact closes on line B. If CR 0317 is then energized by one of its two contacts just after we go past B, CR 0053 will not go on immediately. The CR 0317 contact on line B will therefore not close until we go past to B on the next scan.

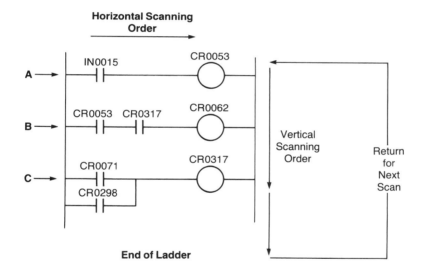

Figure 7–6
PC Scanning Sequence Example

PC OPERATIONAL FAULTS

Every PC has codes for incorrect programming and misoperation. The codes appear on the monitor in code form in small systems or in user-friendly language in larger systems when something is incorrect. For example, "24" might appear on the screen of your small system. Reference to your operations manual code list defines "24" as a memory overflow. In a larger system, the words "memory overflow" (or "illegal key," or whatever the misoperation is) would appear on the screen itself.

In the case of system misconnection, or poor connections, you will get a message such as "communication error." For other problems, different messages will appear on the screen, usually at the bottom.

In case of an internal programming "hang up," a fault light on the CPU will go on. Reference to the operating manual is required for correct interpretation. Typically, a fault light going on at the CPU indicates that a memory-clearing procedure must be carried out. The resetting procedure involves completely clearing the PC program memory. If the program being used has not been previously recorded on tape or disk, a manual keyboard reentry will be required. To prevent the time-consuming manual reentry process, it is a good idea to have each operating program recorded in case a fault occurs. After the clearing procedure, the recorded program can be quickly reentered into the CPU.

FAIL-SAFE CIRCUITS

Some PC circuits are programmed to be turned off by applying a signal voltage. For example, the LATCH–UNLATCH function requires an unlatch signal to turn the coil or output off. If you lose control power, pushing the stop button has no effect and the coil remains on, since control power is needed for system turn-off.

Emergency stop switches or push buttons that are independent of the PC on-off circuits should be included. Figure 7–7 shows a circuit that could

Figure 7–7
Safety "Fail-Safe" Circuit

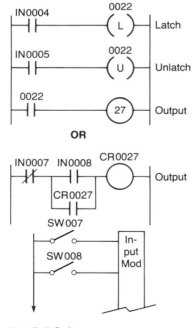

Not Fail Safe
Both require control power available to turn off the output 0027.

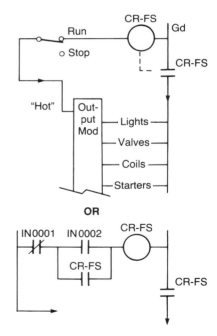

Fail Safe
Circuitry in addition to the PC circuitry is used. Depressing stop switch *or* loss of control power turns outputs off.

be used as a true "fail-safe" system. Turning the master run-stop switch off in the fail-safe circuit on the right deenergizes all coils.

Of course, in the system, "fail-safe" should be defined. You may not wish to turn off all coils when the emergency is pushed. Suppose that a device is spring return. If you expect the emergency switch to stop the machine where it is, it won't; it springs back. For true "fail-safe," a complete control system analysis is needed.

EXERCISES

To carry out chapter exercises, obtain two or more manufacturers' manuals for different PCs. They will be used for reference for exercises 1 through 6.

1. Compare keyboard layouts and functions for two or more PC models. How are they alike? What are their major differences?

2. Compare programming formats for coils and contacts for two or more models. How do the procedure sequences differ? What reference numbers and letters must be used for each system?

3. Compare other function formats such as timers, counters, and sequences in the manner of exercise 2. What are the major differences and similarities?

4. What are the programming ladder arrangement rules for one of the models chosen for analysis? What format arrangements will not be accepted by the CPU?

5. What is the scan rate for the units chosen?

6. What corrective procedures are to be taken when a CPU fault light goes on? If there is more than one fault light, what are the corrective procedures for each?

7. There are four emergency stop buttons on a machine controlled by a PC. Design a "fail-safe" system for the machine.

Registers and Addresses

8

At the end of this chapter, you will be able to

☐ List the five common types of PC registers.
☐ Describe the function of each of the five register types.
☐ Describe how each of the five types of PC registers is used in PC operations.

INTRODUCTION

This chapter covers an important concept for PCs—registers and addresses.

All computers have internal "slots" for storage of data and instructions. In some PCs, the slots are called registers. In other systems they are called addresses. In this text we use the word *register*.

PC registers are of various types and are used in various ways. The value or contents of some PC registers can be changed by moving new values into them from elsewhere in the PC. The register's previous value is erased and lost.

Some other types of PC register values are fixed and cannot be changed. Other PC register values can be altered by inserting new values into them from elsewhere in the PC. The contents of some other types of PC registers can be moved to other PC locations including the PC output modules and terminals. Still other PC registers are used strictly internally where counting, timing, and other functions take place.

This chapter will deal with the five most common types of registers found in PC systems.

GENERAL CHARACTERISTICS OF REGISTERS

Registers can be 4, 8, 16, or 32 bits wide, depending on the PC system you are using. For our illustrations, we will use registers with 16 bits. Each has a value of either 1 or 0.

You can observe register contents on a screen by calling up the register on the keyboard. Additionally, on many models you can print the register contents on a typed printout. Various numbering systems are possible for reading register contents or printing them out. Chapter 9 will describe a variety of PC numbering systems. Depending on your PC capabilities, you may choose to print register values based on one or more different numbering systems. For example, one model allows you to choose between 1—Decimal, 2—Binary, 3—Hex, or 4—ASCII. Other possibilities are Octal, or special codes unique to the system being used. Still other PCs are confined to displaying or printing in only one numbering system, usually the decimal system.

Another important characteristic of PC registers is that they may be manipulated and changed. Chapter 18 will explain how to move new data from one register to another. Other chapters following 18 will explain how registers are changed, analyzed and utilized for various operational PC programs.

Some PCs use prefixes followed by numbers, as we are doing in this chapter. Others have a certain numerical series of addresses assigned to a specific task or function. For example, one model of PC has the addresses 901 through 930 assigned to timers and counters only. Another model might have addresses 31 through 51 assigned as internal registers. It is important to determine the functions for the addresses of your PC registers from its operational manual.

REGISTER DESCRIPTIONS

We will describe each of the five types of registers. Note that all of the registers used in this text for illustration are 16 bits wide.

Holding Registers

A holding register is a type of register in the "middle" of the CPU. It keeps track of the internal workings of the computer processes. It is not directly accessible to inputs or outputs. Refer to your PC user's manual to find out how many holding registers there are in your PC. In small PCs there may be only 16 holding-type registers, or perhaps none are accessible at all. In large machines there are hundreds of holding registers, all accessible for programming use, manipulation, and visual analysis.

Input Registers

The input register has basically the same characteristics as the holding register, except that it is readily accessible to the input module's terminals or ports. The number of input registers in a PC is normally one-tenth that of holding registers.

Output Registers

The output register, like the input register, has the same basic characteristics as the holding register. The output register differs from the input register, however, in that it is readily accessible to the output module's terminals and ports. The number of output registers is normally equal to the number of input registers.

Input Group Registers

The input group register, IG, is somewhat like the input register. It differs in that each one of the individual 16 bits is directly accessible from one input port. One input group register receives data from 16 consecutive input ports. Figure 8-1 illustrates how this IG system works. The advantage of the IG system is that only one register is required to service 16 inputs. Without the IG system, you would need 16 registers to service 16 inputs. Without the input group system, you would use up more computer space to run your programs.

The input module port corresponds to a single input group register bit. Each IG register status controls one bit's status. When a port is enabled, or on, it creates a 1 in the corresponding bit slot. If the port is off, it produces a 0 in the corresponding bit slot.

It is necessary to know how your PC groups the input numbers that correspond to each input group register. A typical scheme is shown in figure 8-2.

Output Group Registers

The output group register, OG, is organized in a manner similar to the input group register. It differs from the IG in a manner similar to the difference between input registers and output registers. Figure 8-3 shows how the OG register functions. One OG register can control 16 outputs. If a 1 is in a bit position, it will turn that bit's corresponding output on. A 0 will turn its corresponding output off. The grouping scheme for output group register is similar to the input group register system. The grouping scheme is shown in figure 8-3. The output group register is particularly useful in sequencer operation, as we shall see in chapter 20.

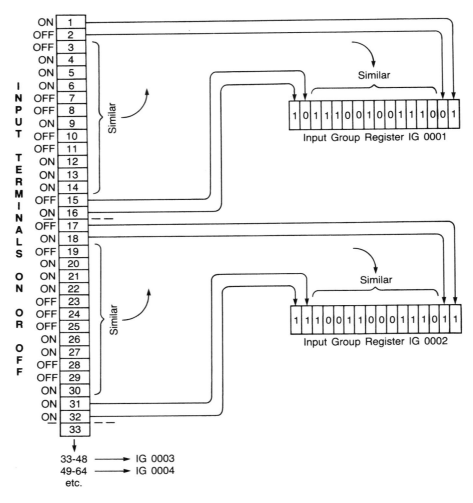

Figure 8-1
Input Group
Register Scheme

Figure 8-2
Input Group/
Input Port
Numbering
Scheme

Input Group Register Number	8 Bit System - Inputs Controlled	16 Bit System - Inputs Controlled
1	1-8	1-16
2	9-16	17-32
3	17-40	33-48
4	41-48	49-64
5	49-56	65-80
6	57-64	81-96
7 etc.	65-72 etc.	97-102 etc.

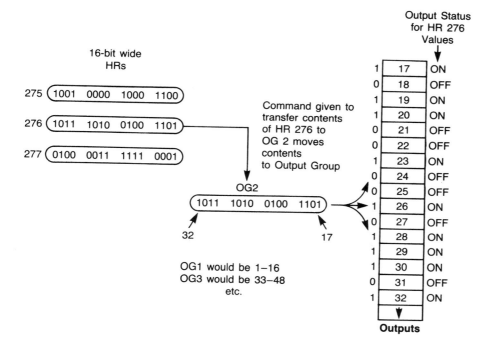

Output Status
for HR 276
Values

Figure 8-3
Output Group
Register Scheme

16-bit wide
HRs

275 (1001 0000 1000 1100)

276 (1011 1010 0100 1101)

277 (0100 0011 1111 0001)

Command given to
transfer contents
of HR 276 to
OG 2 moves
contents
to Output Group

OG2
(1011 1010 0100 1101)

32   17

OG1 would be 1–16
OG3 would be 33–48
etc.

1	17	ON
0	18	OFF
1	19	ON
1	20	ON
0	21	OFF
0	22	OFF
1	23	ON
0	24	OFF
0	25	OFF
1	26	ON
0	27	OFF
1	28	ON
1	29	ON
1	30	ON
0	31	OFF
1	32	ON

Outputs

EXERCISES

1. List the five major types of registers. Use a block diagram to show where each type fits into the PC scheme of operation.
2. What would the input group registers look like for the three input module status arrangements shown in figure 8–4? What would the

Figure 8-4
Diagram for
Exercise 2

A		B		C	
Input No.	Status	Input No.	Status	Input No.	Status
49	ON	105	ON	209	OFF
50	ON	106	OFF	210	ON
51	OFF	107	ON	211	OFF
52	ON	108	OFF	212	ON
53	OFF	109	ON	213	ON
54	OFF	110	OFF	214	OFF
55	ON	111	OFF	215	ON
56	ON	112	ON	216	OFF
57	ON			217	ON
58	OFF	(8 Bit PC)		218	ON
59	ON			219	OFF
60	OFF			220	OFF
61	ON			221	OFF
62	ON			222	ON
63	OFF			223	ON
64	ON			224	OFF

number of each IG register be? What would the register contents be, in binary?

3. What would be the status of the corresponding outputs for the four output and input group registers shown in figure 8–5? What are the corresponding output numbers for the four OG and IG registers shown?

Figure 8–5
Diagram for
Exercise 3

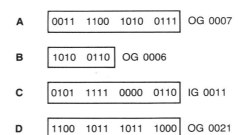

A   `0011  1100  1010  0111`   OG 0007

B   `1010  0110`   OG 0006

C   `0101  1111  0000  0110`   IG 0011

D   `1100  1011  1011  1000`   OG 0021

Number Conversions

9

At the end of this chapter, you will be able to

☐ Explain the basis of the binary counting system.
☐ Convert numbers from binary to decimal and vice versa.
☐ Explain the basis of the BCD system.
☐ Convert numbers from BCD to binary and vice versa.
☐ Program a PC for binary-to-BCD and BCD-to-binary conversions.
☐ Explain the basis of the octal and hex systems.
☐ Convert among the three systems—decimal, octal, and hex.
☐ Describe the three code systems—Gray, ASCII, and EBCDIC.

INTRODUCTION

Most PC functional and programming operations can be handled in the decimal numbering system. Some PC models and individual PC functions use other numbering systems. This chapter deals with some of these numbering systems, including binary, BCD, octal, hexadecimal, Gray, ASCII, and EBCDIC. The basics of each system will be explained and conversions from one system to another will be illustrated.

If your PC works entirely in decimal, this chapter is optional. The purpose of the chapter is to give you enough background to handle any numbering system that you may encounter on larger PCs or in the future. Various texts are available that detail the use of all numbering systems.

Note that this chapter does all conversions "long hand." Actual conversions can be made easily with a moderately priced, scientific, hand-held calculator. Calculators normally handle only up to 511 decimal. There are various personal computer programs available for larger numbers.

THE BINARY SYSTEM

This section describes the binary numbering system, how it differs from decimal, and how to convert from one to the other.

The decimal numbering system, which uses a base of 10, is the one we use every day. The decimal system is said to have developed by tens because the originators had ten fingers and so could easily count to ten. Above ten, another person or means had to be used to keep track of the tens. If the count was over 99 (and less than 1000) a third means was needed to keep track of the 100s. Thus, the decimal system was developed.

When digital computers were developed, they were designed to count by 2's; their counters were either off or on. Therefore, the base of the computer counting system is 2, instead of the 10 used in the decimal system. This base-two binary system, which had been around for a long time, became the basis for all computer systems. With the advent of Integrated Circuit (IC) chips, hundreds of binary, on-off switches can now be found on a single IC chip in computers. All PCs work internally in the binary system since they are IC-chip based. Since outside information is decimal or BCD (BCD is defined later in the chapter), a conversion is needed to and from the binary used in the PC CPU.

Figure 9–1 shows a comparison of the interpretation of a decimal and a binary number. The numbers are arbitrarily picked for this illustration. There are 10 possible count values (0 through 9) in each decimal position. For the binary counter, there are only 2 possible counts, 0 or 1. As a result, for binary we must move to the next bit or "slot" after we reach 1, not 9, as in decimal. The binary number shown has a decimal equivalent value of 4 + 1, or 5.

Figure 9–1
Decimal and
Binary System
Comparison

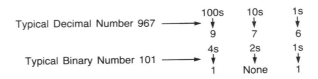

Figure 9–2 compares the value of a given quantity for decimal and binary systems.

Figure 9–3 illustrates the counting "slots" or bit status for some typical binary numbers. The more bits you use, the higher you may count. The maximum decimal equivalent count is given for each example (1 in all bits). The equivalent decimal values for the given binary bit patterns are given also. Typical PCs work in 8- or 16-bit form. Other PCs may have 4, 6, 12, or other numbers of bits as a base for each address or register.

Figure 9–4 gives some examples of conversions from binary to decimal. Four-, 8-, and 12-bit numbers are shown in the example. Different numbers of bits are converted similarly, starting from right to left.

Figure 9–5 illustrates the opposite conversion, from decimal to binary. The conversion is not as easily accomplished as binary to decimal, especially as the numbers get larger. Basically, the procedure is to find the largest number in the digital columns that is less than the decimal number being converted. A 1 is placed in that binary column. Next, subtract the

Decimal-Base 10	Binary-Base 2
1	1
2	10
3	11
4	100
5	101
6	110
7	111
8	1000
9	1001
10	1010
11	1011
12	1100
13	1101
14	1110
15	1111
16	10000
17	10001
18	10010
19	10011
20	10100
21	10101
22	10110
23	10111
24	11000
25	11001

Figure 9-2
Decimal and Binary Value Comparison

Bits	Maximum Value - IF All Bits Are 1	Typical Binary Number															Equivalent Decimal Value
4	15	8s 1	4s 1	2s 0	1s 1												= 13
8	255	128s 1	64s 0	32s 0	16s 1	8s 1	4s 1	2s 0	1s 1								= 157
16	4095	2048s 1	1024s 0	512s 1	256s 1	128s 0	64s 1	32s 1	16s 0	8s 0	4s 1	2s 0	1s 1				= 2917
32	65,553	32786 1	16384 0	8192 ?	4096 0	2048 1	1024 1	512 0	256 0	128 1	64 1	32 1	16 0	8 1	4 1	2 1	1 0	= 44,288

handwritten notes: 44288 11502   11502 2310 larger. less less larger   11502 ↑more than 16384

Figure 9-3
Typical Binary Numbers and Decimal Equivalents

Digital Number				Conversion	Decimal Result
			1011	(8 + 0 + 2 + 1)	11
		0101	0101	(0 + 64 + 0 + 16) + (0 + 4 + 0 + 1)	85
		0011	1110	(0 + 0 + 32 + 16) + (8 + 4 + 2 + 0)	62
		1101	0011	(128 + 64 + 0 + 16) + (0 + 0 + 2 + 1)	211
1010		1010	1111	(2048 + 0 + 512 + 0) + (128 + 0 + 32 + 0) + (8 + 4 + 2 + 1)	2735

Figure 9-4
Binary-to-Decimal Conversions

Figure 9–5
Decimal-to-
Binary Con-
version

Decimal Number ⟶　　　　　　　Binary Number

Decimal Number	128	64	32	16	8	4	2	1
1	0	0	0	0	0	0	0	0
3	0	0	0	0	0	0	1	1
7	0	0	0	0	0	1	1	1
22	0	0	0	1 (22 − 16 = 6)	0	1 (6 − 4 = 2)	1	0
87	0	1 (87 − 64 = 23)	0	1 (23 − 16 = 7)	0	1	1	1
125	0	1 (125 − 64 = 61)	1 (61 − 32 = 29)	1 (29 − 16 = 13)	1 (13 − 8 = 5)	1 (5 − 4 = 1)	0	1
164	1 (164 − 128 = 36)	0	1 (36 − 32 = 4)	0	0	1	0	0

column's binary/decimal value number from the decimal number being converted. Then move one column to the right and apply the same procedure to the remainder. Keep repeating the procedure until the remainder is zero.

BINARY CODED DECIMAL NUMBERING SYSTEM

Another related number system used often in PCs is the Binary Coded Decimal, or BCD, system.

The key to the BCD system is to utilize four digital bits for each single decimal number of output or input. A commonly used BCD system contains four decimal numbers and can display decimal values from 0000 through 9999. The four-digit decimal number is then represented by 4 times 4, or 16 bits. Figure 9–6 shows some typical decimal numbers and their BCD equivalents. For comparison, the binary values are also shown in the table.

PC CPUs function in binary, not BCD or decimal. If the PC received a BCD number from an input thumbwheel (a small rotary device set to 0 through 9 by rotating it with the thumb or finger), it would interpret the number as a binary number. The BCD input number must be converted to its binary equivalent for correct PC CPU operation. We could convert the number manually, as shown in Figure 9–6; however, if it must be done 25 times per second, for example, manual conversion would be impossible. For fast operation, the conversion must be performed by the PC. A PC function for the conversions is available. Your PC may do the conversion automatically; however, in most cases, you must program in the BCD-to-binary conversion for inputs. For outputs, you program binary-to-BCD conversions.

Figure 9–7 shows a simple example of a required conversion. The PC is required to take an input value in BCD and multiply it by 0.5. The re-

Decimal	BCD		Binary (Ref.)
1		0001	1
2		0010	10
3		0011	11
4		0100	100
5		0101	101
6		0110	110
7		0111	111
8		1000	1000
9		1001	1001
10	0001	0000	1010
11	0001	0001	1011
12	0001	0010	1100
13	0001	0011	1101
14	0001	0100	1110
15	0001	0101	1111
16	0001	0110	10000
17	0001	0111	10001
18	0001	1000	10010
19	0001	1001	10011
20	0010	0000	10100
21	0010	0001	10101
22	0010	0010	10110
23	0010	0011	10111
24	0010	0100	11000
25	0010	0101	11001

Figure 9-6
Decimal and
BCD Equivalents

Table of Values

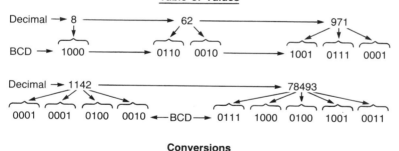

Conversions

sulting number is then displayed in BCD on an output LED readout. Since the PC functions in binary, the input BCD value must be converted to binary before internal processing can take place. Then, the converted binary number is multiplied in binary by 0.5. This resulting binary value is then converted to BCD for output to the display.

PC Conversions Between Decimal and BCD

This section of the chapter will show the formats of the PC number conversion functions and will use the conversion functions in a PC program. Many PCs include these functions for converting BCD coded numbers

Figure 9-7
BCD-to-Binary
and Binary-
to-BCD PC
Processing

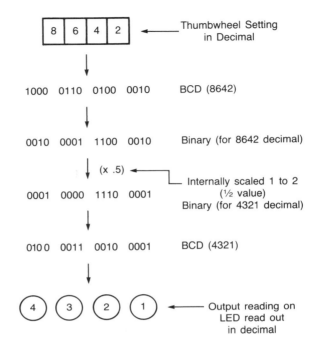

to binary for use by the CPU. They also can contain the inverse conversion, binary to BCD. Figure 9–8 is a typical PC application problem requiring number conversions. It illustrates in block diagram form how the conversions are accomplished. The mathematical manipulation in the middle block can take any form.

Figure 9–9 shows the layout of a typical BCD-to-binary conversion function, which is usually used for input data conversions. It converts the BCD value found in the source register to a binary number in the destination register for PC CPU use. When the function's input line is energized, values in the source register are converted from BCD to binary. The resulting number is put into the destination register. Typically, the coil comes on only if one of the BCD digits exceeds 9 during mathematical conversion.

Figure 9–10 shows how the reverse of the conversion in figure 9–9 is accomplished. It converts a binary value in the source to a BCD value in the destination. This function's primary use is for feeding an output display. Typically for this function, the coil comes on only if the binary value exceeds 9999 in decimal during operation.

To learn how the individual functions of figures 9–9 and 9–10 function, use the MONITOR mode. Program the function to be observed in the EDIT mode, place the PC in the MONITOR mode, and then call up the

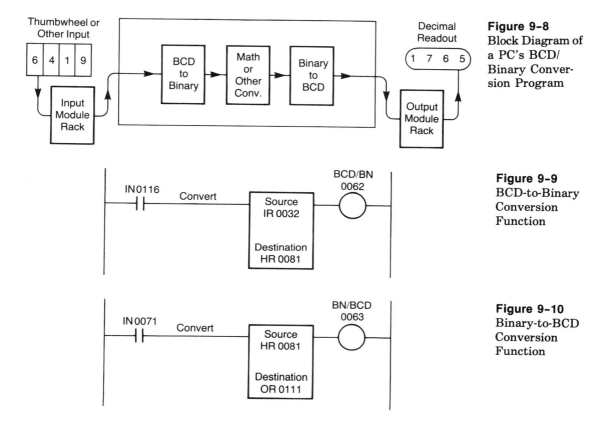

Figure 9-8
Block Diagram of a PC's BCD/ Binary Conversion Program

Figure 9-9
BCD-to-Binary Conversion Function

Figure 9-10
Binary-to-BCD Conversion Function

two registers involved in the conversion function. Insert appropriate values into the first register. Enable the conversion function being used and observe the resulting value in the second register. Verify the conversion function's correct mathematical operation manually or by calculator conversion computation.

THE OCTAL AND HEX CODES

Two other number codes often encountered in dealing with PCs and computers are the octal code and the hexadecimal, or hex, code. The octal code uses a base of 8 and the hex uses a base of 16. These bases are in contrast to decimal and BCD, which have a base of 10. If the computer's numbers are three bits long, it is in the octal numbering system. Three bits can count up to 7 as a maximum and start over at 8. If four bits are used for each number, it is in hex. Four bits can count up to 15 and then start over at 16. Figure 9-11 shows the comparison of octal- and hex-to-decimal equivalents.

Figure 9–11
Decimal—Octal—
Hex System
Comparison

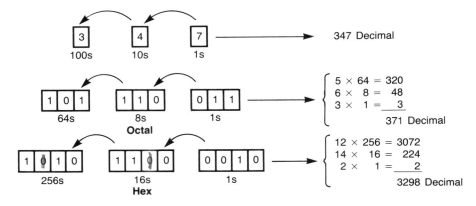

Figure 9–12 gives a table comparing values for decimal numbers versus their octal equivalents (and binary for reference). When we reach 8, octal starts over and a 1 goes in the slot to the left.

How do we convert between octal and decimal? Figure 9–13 shows some typical octal-to-decimal conversions and figure 9–14 illustrates some decimal-to-octal conversions. For decimal-to-octal conversions, repetitive division is used.

Figure 9–12
Octal Code Table

Decimal	Octal	(Ref.) Binary
Base (10)	Base (8)	Base (2)
1	1	1
2	2	10
3	3	11
4	4	100
5	5	101
6	6	110
7	7	111
8	10	1000
9	11	1001
10	12	1010
11	13	1011
12	14	1100
13	15	1101
14	16	1110
15	17	1111
16	20	10000
17	21	10001
18	22	10010
19	23	10011
20	24	10100
21	25	10101
22	26	10110
23	27	10111
24	30	11000
25	31	11001

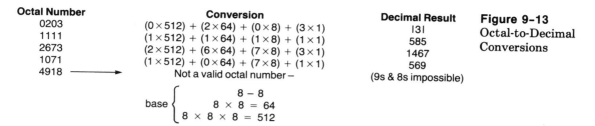

Octal Number	Conversion	Decimal Result	Figure 9-13
0203	$(0 \times 512) + (2 \times 64) + (0 \times 8) + (3 \times 1)$	131	Octal-to-Decimal
1111	$(1 \times 512) + (1 \times 64) + (1 \times 8) + (1 \times 1)$	585	Conversions
2673	$(2 \times 512) + (6 \times 64) + (7 \times 8) + (3 \times 1)$	1467	
1071	$(1 \times 512) + (0 \times 64) + (7 \times 8) + (1 \times 1)$	569	
4918 $\longrightarrow$	Not a valid octal number –	(9s & 8s impossible)	

$$\text{base} \begin{cases} 8 - 8 \\ 8 \times 8 = 64 \\ 8 \times 8 \times 8 = 512 \end{cases}$$

Decimal Number	Conversion	Octal Number	Figure 9-14
5	None Needed	5	Decimal-to-Octal
11	$11/8 = 1$ with 3 Remainder	13	Conversions
28	$28/8 = 3$ with 4 Remainder	34	
85	$85/8 = 10$ with 5 Remainder		
	↳$10/8 = 1$ with 2 Remainder	125	
116	$116/8 = 14$ with 4 Remainder		
	↳$14/8 = 1$ with 6 Remainder	164	
982	$982/8 = 122$ with 6 Remainder		
	↳$122/8 = 15$ with 2 Remainder	1726	
	↳1 with 7 Remainder		

Figure 9-15
Decimal—Hex—Binary Comparison

Decimal Base 10	Hex Base 16	Binary Base 2
0	0	0000
1	1	0001
2	2	0010
3	3	0011
4	4	0100
5	5	0101
6	6	0110
7	7	0111
8	8	1000
9	9	1001
10	A	1010
11	B	1011
12	C	1100
13	D	1101
14	E	1110
15	F	1111
16	10	10000
17	11	10001

As previously stated, the hex system uses 4 bits per number, compared to octal's 3. Therefore, hex is base 16, as was shown in figure 9-11. Since we must count beyond 9 in a single column, we use sequential letters of the alphabet for numbers 10 through 15. The hex system is shown in figure 9-16.

Figure 9-16 shows how hex numbers are converted to decimal. The conversion uses straightforward multiplication, as does the octal-to-decimal conversion.

Figure 9-16
Hex-to-Decimal
Conversion

Hex Number	Conversion	Decimal Number
13	$(1 \times 16) + (3 \times 1)$	19
BC	$(11 \times 16) + (12 \times 1)$	188
F4D	$(15 \times 256) + (4 \times 16) + (13 \times 1)$	3917
C1B7	$(12 \times 4096) + (1 \times 256) + (11 \times 16) + (7 \times 1)$	49591

$$\text{Base} \begin{cases} 16 \\ 16 \times 16 = 256 \\ 16 \times 16 \times 16 = 4096 \end{cases}$$

Figure 9-17
Decimal-to-Hex
Conversions

Decimal Numbers	Conversion	Hex Number
7	—	7
12	—	C
21	21/16 = 1 with 5 Remainder	15
111	111/16 = 6 with 15 Remainder	6F
247	247/16 = 15 with 7 Remainder	F7
398	398/16 = 24 with 14 Remainder	18E
	↳ 24/16 = 1 with 8 Remainder	

The conversion of decimal to hex is a little more involved than hex to decimal. We use repetitive division similar to that used for decimal-to-octal conversions, as shown in figure 9–17.

THREE OTHER CODE SYSTEMS

There are three other codes often encountered in PC work. These codes are the Gray code, the ASCII code, and the EBCDIC code. The Gray code involves only numbers; ASCII and EBCDIC codes involve both numbers and letters or symbols.

The Gray code's structure, which is shown in figure 9–18, is constructed so that only one digit changes as you go down one step. You will note this arrangement by following down the left Gray code column. Note the step from the fourth step to the fifth step. Only one number changes for the Gray code. The third digit changes from a 0 to a 1. For the same binary step, three bit's digits change.

The single-digit change is important when certain mechanical and photo-electric encoders are used. Details of these encoding systems may be found in various texts on control systems. If you follow the binary code in the right hand column, you will see that one, two, three, or more digits change from one number to the next. The Gray code has only one change per step. A change by more than one digit is difficult to handle when using encoders.

The Gray code and standard binary are different and are not inter-changeable. Caution must be exercised not to mix the two. If your PCs works in binary, putting Gray code values into it will give false input information. The inverse is also true for output; a Gray code output will not function properly in binary.

Gray Code	Binary
0000	0000
0001	0001
0011	0010
0010	0011
0110	0100
0111	0101
0101	0110
0100	0111
1100	1000
1101	1001
1111	1010
1110	1011
1010	1100
1011	1101
1001	1110
1000	1111

Figure 9-18
The Gray Code
Compared to the
Binary Code

ASCII stands for American Standard Code for Information Interchange. As you can see from its listing in figure 9–19, it covers numbers, letters, symbols, and abbreviations. The ASCII code requires six or seven memory bits, depending on the system used. It is used to interface the PC CPU with alphanumeric keyboards and printers.

The EBCDIC code is similar to the ASCII code in function. It is shown in figure 9–20. EBCDIC uses eight bits, compared to the six or seven bits used for the ASCII code. Whether ASCII or EBCDIC is used depends on the particular PC's operational requirements and capabilities.

Hexadecimal	Decimal	Octal	Binary	Character	Description
00	0	000	0000000	NUL	Null
01	1	001	0000001	SOH	Start of heading
02	2	002	0000010	STX	Start of text
03	3	003	0000011	ETX	End of text
04	4	004	0000100	EOT	End of transmission
05	5	005	0000101	ENQ	Enquiry
06	6	006	0000110	ACK	Acknowledge
07	7	007	0000111	BEL	Bell
08	8	010	0001000	BS	Back space
09	9	011	0001001	HT	Horizontal tab
0A	10	012	0001010	LF	Line feed
0B	11	013	0001011	VT	Vertical tab
0C	12	014	0001100	FF	Form feed
0D	13	015	0001101	CR	Carriage return
0E	14	016	0001110	SO	Shift out
0F	15	017	0001111	SI	Shift in
10	16	020	0010000	DLE	Data link escape
11	17	021	0010001	DC1	Device Control 1
12	18	022	0010010	DC2	Device Control 2
13	19	023	0010011	DC3	Device Control 3

Figure 9-19
The ASCII Code

Hexadecimal	Decimal	Octal	Binary	Character	Description
14	20	024	0010100	DC4	Device Control 4
15	21	025	0010101	NAK	Negative acknowledge
16	22	026	0010110	SYN	Synchronize
17	23	027	0010111	ETB	End of transmission block
18	24	030	0011000	CAN	Cancel
19	25	031	0011001	EM	End of media
1A	26	032	0011010	SUB	Substitute
1B	27	033	0011011	ESC	Escape
1C	28	034	0011100	FS	File separator
1D	29	035	0011101	GS	Group separator
1E	30	036	0011110	RS	Record separator
1F	31	037	0011111	US	Unit separator
20	32	040	0100000	SP	Space
21	33	041	0100001	!	Exclamation
22	34	042	0100010	''	Double quote
23	35	043	0100011	#	Number or pound
24	36	044	0100100	$	Dollar sign
25	37	045	0100101	%	Percentage
26	38	046	0100110	&	Ampersand
27	39	047	0100111	'	Apostrophe or single quote
28	40	050	0101000	(	Left parenthesis
29	41	051	0101001	)	Right parenthesis
2A	42	052	0101010	*	Asterisk
2B	43	053	0101011	+	Plus
2C	44	054	0101100	,	Comma
2D	45	055	0101101	−	Minus
2E	46	056	0101110	.	Period
2F	47	057	0101111	/	Slash
30	48	060	0110000	0	Zero
31	49	061	0110001	1	One
32	50	062	0110010	2	Two
33	51	063	0110011	3	Three
34	52	064	0110100	4	Four
35	53	065	0110101	5	Five
36	54	066	0110110	6	Six
37	55	067	0110111	7	Seven
38	56	070	0111000	8	Eight
39	57	071	0111001	9	Nine
3A	58	072	0111010	:	Colon
3B	59	073	0111011	;	Semi-colon
3C	60	074	0111100	<	Less than
3D	61	075	0111101	=	Equal
3E	62	076	0111110	>	Greater than
3F	63	077	0111111	?	Question
40	64	100	1000000	@	At sign
41	65	101	1000001	A	Letter A
42	66	102	1000010	B	Letter B
43	67	103	1000011	C	Letter C
44	68	104	1000100	D	Letter D
45	69	105	1000101	E	Letter E
46	70	106	1000110	F	Letter F
47	71	107	1000111	G	Letter G
48	72	110	1001000	H	Letter H

Hexadecimal	Decimal	Octal	Binary	Character	Description
49	73	111	1001001	I	Letter I
4A	74	112	1001010	J	Letter J
4B	75	113	1001011	K	Letter K
4C	76	114	1001100	L	Letter L
4D	77	115	1001101	M	Letter M
4E	78	116	1001110	N	Letter N
4F	79	117	1001111	O	Letter O
50	80	120	1010000	P	Letter P
51	81	121	1010001	Q	Letter Q
52	82	122	1010010	R	Letter R
53	83	123	1010011	S	Letter S
54	84	124	1010100	T	Letter T
55	85	125	1010101	U	Letter U
56	86	126	1010110	V	Letter V
57	87	127	1010111	W	Letter W
58	88	130	1011000	X	Letter X
59	89	131	1011001	Y	Letter Y
5A	90	132	1011010	Z	Letter Z
5B	91	133	1011011	[	Left bracket
5C	92	134	1011100	\	Back slash
5D	93	135	1011101	]	Right bracket
5E	94	136	1011110	↑	Up arrow
5F	95	137	1011111	←	Back arrow
60	96	140	1100000	'	Back quote or accent mark
61	97	141	1100001	a	Small letter a
62	98	142	1100010	b	Small letter b
63	99	143	1100011	c	Small letter c
64	100	144	1100100	d	Small letter d
65	101	145	1100101	e	Small letter e
66	102	146	1100110	f	Small letter f
67	103	147	1100111	g	Small letter g
68	104	150	1101000	h	Small letter h
69	105	151	1101001	i	Small letter i
6A	106	152	1101010	j	Small letter j
6B	107	153	1101011	k	Small letter k
6C	108	154	1101100	l	Small letter l
6D	109	155	1101101	m	Small letter m
6E	110	156	1101110	n	Small letter n
6F	111	157	1101111	o	Small letter o
70	112	160	1110000	p	Small letter p
71	113	161	1110001	q	Small letter q
72	114	162	1110010	r	Small letter r
73	115	163	1110011	s	Small letter s
74	116	164	1110100	t	Small letter t
75	117	165	1110101	u	Small letter u
76	118	166	1110110	v	Small letter v
77	119	167	1110111	w	Small letter w
78	120	170	1111000	x	Small letter x
79	121	171	1111001	y	Small letter y
7A	122	172	1111010	z	Small letter z
7B	123	173	1111011	{	Left brace
7C	124	174	1111100	|	Vertical bar
7D	125	175	1111101	}	Right brace
7E	126	176	1111110	∿	Approximate or tilde
7F	127	177	1111111	DEL	Delete (rub out)

Figure 9–20
The EBCDIC
Code

Character	ASCII	EBCDIC
0	011 0000	1111 0000
1	011 0001	1111 0001
2	011 0010	1111 0010
3	011 0011	1111 0011
4	011 0100	1111 0100
5	011 0101	1111 0101
6	011 0110	1111 0110
7	011 0111	1111 0111
8	011 1000	1111 1000
9	011 1001	1111 1001
A	100 0001	1100 0001
B	100 0010	1100 0010
C	100 0011	1100 0011
D	100 0100	1100 0100
E	100 0101	1100 0101
F	100 0110	1100 0110
G	100 0111	1100 0111
H	100 1000	1100 1000
I	100 1001	1100 1001
J	100 1010	1101 0001
K	100 1011	1101 0010
L	100 1100	1101 0011
M	100 1101	1101 0100
N	100 1110	1101 0101
O	100 1111	1101 0110
P	101 0000	1101 0111
Q	101 0001	1101 1000
R	101 0010	1101 1001
S	101 0011	1110 0010
T	101 0100	1110 0011
U	101 0101	1110 0100
V	101 0110	1110 0101
W	101 0111	1110 0110
X	101 1000	1110 0111
Y	101 1001	1110 1000
Z	101 1010	1110 1001

EXERCISES

Convert the given numbers from one system to the other specified system
for exercises 1 through 8.

1. Binary to decimal:
   111, 1011, 10 1010, 110 1101, 1011 0011, 1001 1011 1110
2. Decimal to binary:
   14, 42, 85, 162, 207, 459
3. Decimal to BCD:
   56, 381, 1121, 4583, 6666

4. BCD to decimal:
   0110 0011 1011 1100 (Hint: is this possible?)
   0110 0011 0111 0110
   1000 1001 0100 0001
   0101 0000 1000 0101
   0011 0011 0011 0111

5. Decimal to octal:
   7, 15, 88, 327, 691, 1121

6. Octal to decimal:
   6, 35, 77, 201, 847, 4464

7. Decimal to hex:
   8, 14, 79, 410, 558, 1243

8. Hex to decimal:
   11, 3C, 2A2, BCF, 1B4C, DDDDD

9. Octal uses three bits and hex uses four bits. What would a quad number system look like, using only two bits? Construct a table of comparison values to decimal.

Relation of Digital Gate Logic to Contact Logic

10

At the end of this chapter, you will be able to

☐ List the six basic digital gate types, draw their symbols, and describe their function.
☐ Show the relation of switch contact logic (relay/PC ladder logic) to digital logic for each gate type.
☐ Create digital systems and PC/relay logic digrams from process word descriptions.
☐ Convert from any one of the three programming systems to any other for:
  Process operation word description.
  Relay logic diagrams/PC logic diagrams.
  Digital gate diagrams.
☐ Write Boolean expression.

INTRODUCTION

Large PC programming systems of the screen/monitor type do not require the use of digital gate logic principles. The programming is normally done by typing in lines, connection nodes, contacts, and coils or functions. However, most smaller programmers with smaller LED displays have keyboard keys with digital logic notations.

These smaller programmers can have digital gate logic keys such as "AND," "OR," "NOT," and others. Chapter 10 will show how to relate these logic terms to the relay and large-screen PC logic. Once the logic terms are understood, PC programming using them can be easily accomplished.

Another series of symbols that appear on some PC keyboards are a dot, +, −, 0, and =. These are Boolean algebra symbols. Boolean algebra is a shorthand way of writing digital gate diagrams. Since this type of programming format is not found often, Boolean principles will be discussed only briefly.

There is another reason for studying digital programming. Some computer-trained persons understand PC programming best using digital logic. This chapter will help these people to program PCs properly.

Chapter 10 compares word descriptions, relay/PC ladder diagrams, and digital gate diagrams, and, more important, shows how to translate from one of the three systems to another. The use of a Boolean system and its digital relation will also be reviewed briefly. A person who knows one system but must program in another will be able to do so by mastering this chapter's principles. You may be interested in delving deeper into digital gates and Boolean equivalents. If so, there are many digital logic texts which cover them in detail.

By necessity, the chapter has a high ratio of illustrations to text to fully explain the principles involved.

DIGITAL LOGIC GATES

We will discuss digital logic gates from a PC logic standpoint, but we will not cover the details of their electronic internal workings or their electrical operation. Figure 10-1 shows the six basic types of logic gates.

Figure 10-1
The Six Basic
Digital Logic
Gates

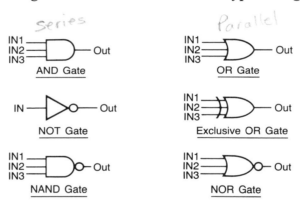

All gates have one output. They are either on (1) or off (0), depending on the logic status of their inputs, on (1) or off (0). A gate-On condition is typical when +5 volts DC come from the output terminal. Off is typically 0 volts output.

The NOT gate always has one input. The EXCLUSIVE OR gate always has two inputs. The other four types can have two to eight inputs and sometimes more. An input on is typical when +5 volts DC are applied to an input terminal. Off is typically 0 volts applied to an input terminal.

There is internal electronic circuitry that causes the gates to function properly. There are usually four gates of one kind on the digital type of integrated circuit chip. The chip, as well as a single gate, requires two additional terminals in addition to its 12 logic terminals. These added two are for a power supply of 5 volts to furnish operational power to the gate or IC chip.

Operation of the Six Digital Gates

The AND gate and its programming equivalents are shown in figure 10-2. For the AND gate output to the on (1), all inputs must be on (1). The relay programming and PC programming equivalents are also shown in the figure. For the four-input situation, input 1, input 2, input 3, and input 4 must be on for output 12 to be on. Otherwise, output 12 is off.

For AND gate programming with a digital PC keyboard, the sequence of key operation would be: 1, and, 2, and, 3, and, 4, =, 12.

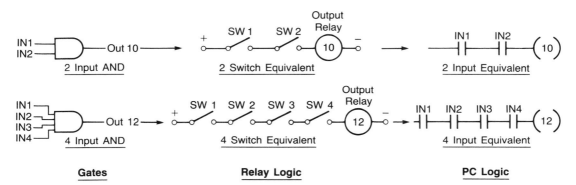

Figure 10-2
The AND Gate and Relay and PC Equivalents

The OR gate operation is shown in figure 10-3. For an OR gate output to be on (1), any one or more of the inputs must be on (1). For the output to be off (0), all inputs must be off (0). The same operational voltages and principles apply to OR as for AND gates.

Again, the equivalent relay and PC programming diagrams are shown in the figure. The word description of this operation is: for output 17 to be on (1), any one or more of inputs 1, 2, and 3 must be on (1); otherwise, output 17 is off (0).

To input an OR keyboard program for figure 10-3, we would use the sequence: 1, or, 2, or, 3, or, =, 17.

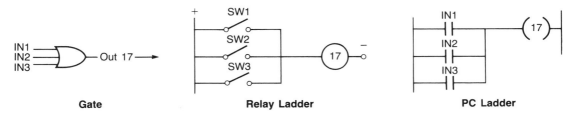

Figure 10-3
The OR Gate and Relay and PC Equivalents

The NOT gate is shown in figure 10-4 along with relay and PC equivalents. It reverses the input logic status, on or off, from the input to the CPU. The output will then be the reverse of the input, off or on. A NOT key input, inserted at the proper point in the program sequence, carries out the reversal.

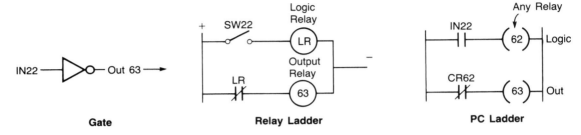

Gate   **Relay Ladder**   **PC Ladder**

Figure 10-4
The NOT Gate and Relay and PC Equivalents

The EXCLUSIVE OR gate symbol was included in figure 10-1. Its output is on (1) when one, and only one, of its two inputs is on (1). If both inputs are on, the output is off. This type of gate is seldom used in PCs, so we will not discuss it beyond showing its basic symbol.

The NAND and NOR gates are the two final basic gates. Both are a combination of two other basic gates. The NAND gate, shown in figure 10-5, is the combination of an AND and a NOT gate. The relay and PC logic for the NAND gate require a logic relay, as shown. The NAND keyboard program would be: 1, nand, 2, nand, 3, nand, 4, =, 27.

Figure 10-5
The NAND Gate
and Relay and
PC Equivalents

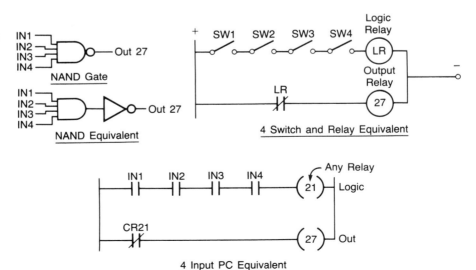

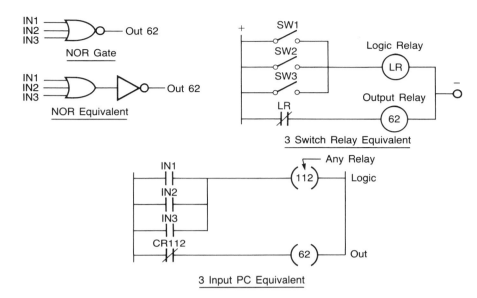

Figure 10-6
The NOR Gate
and Relay and
PC Equivalents

The NOR gate, shown in figure 10-6, is made up of an OR gate and a NOT gate. Again, a logic relay is required for relay or PC logic. The NOR keyboard program would be: 1, nor, 2, nor, 3, nor, 4, =, 62.

BOOLEAN ALGEBRA PC PROGRAMMING

Sometimes you may have to program a PC in the Boolean algebra system, which is a "shorthand" method of writing gate diagrams. Analysis of complex gate diagrams can be easily made when written in Boolean form. The analysis is covered in digital logic texts. This section covers only the PC programming aspects of Boolean algebra.

The symbols used in the Boolean algebra system are illustrated in figure 10-7. Examples of usage and the meaning of the Boolean expression in

Symbol	Definition	Example of Usage	Meaning-Word Description
•	and	$C \cdot D \cdot E$	C and D and E
+	or	$11 + 12$	11 or 12
−	not	$\overline{M}$	Not M
o	invert		Change
=	results in	$F \cdot G = L$	L is true (on) if both F and G are true (on)

Figure 10-7
Boolean Algebra
Symbol
Notation

Figure 10-8
Boolean Algebra
Equivalents for
Digital Gates

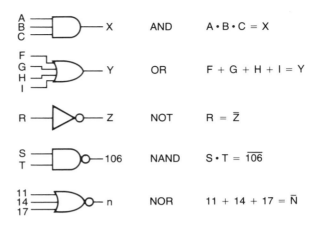

words are also given. Figure 10–8 shows some typical gates and how they would be represented in Boolean form.

CHAPTER EXAMPLES

The ten chapter examples which follow will illustrate conversion from one PC programming system to another. The Boolean expressions are included for reference in each example as optional information. The ten examples are divided into three groups as follows:

A. Four examples of how to convert a word description into ladder and gate diagrams.

B. Three examples of conversion from a ladder diagram to a gate diagram.

C. Three examples of conversion from a gate diagram to a ladder diagram.

Ladder diagrams will be included for relay logic and PC logic.
Example A1 is shown in figure 10–9. Output 122 is to be on only when

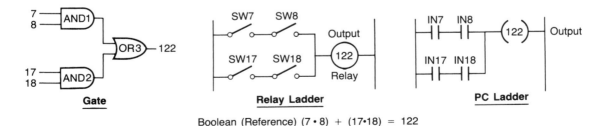

Figure 10-9
Example A1: Word Description Conversion

either inputs 7 and 8 are on or if inputs 17 and 18 are on. Output 122 can be on when all four inputs are on.

A conveyor control problem is given for example A2. Conveyor C is to run when any one of four inputs are on. It is to stop when any one of four other inputs are on. The ladder relay and PC control diagrams are shown.

An explanation of the A2-derived gate diagram is in order. The four starts are inputted to an OR gate, OR 1. When any one of the four is depressed, the OR gate output goes on. When the start is released, the output goes off. The stop buttons are connected to another OR gate, OR 2. OR 2 goes on when any one or more stop buttons are depressed. The OR 2 output is inverted and sent to the AND gate.

With no stop button depressed, the stop OR 2 gate is off (or low). Due to the NOT inversion, the stop input to the AND gate is on (or high) when no stop buttons are depressed. With no stops depressed, output C can go on if OR 1 is also on. Any time that both OR 1 and OR 2 are high, AND 4 is high and output C is on. Example A2 is illustrated in figure 10–10.

Example A3 is a motor control circuit with two start and two stop buttons. When the start button is depressed, the motor runs. By sealing, it continues to run when the start button is released. The stop buttons stop the motor when depressed. This example differs from the previous one in that the system seals on when the start is released.

Example A4 is a more complex system. Its solutions are shown in figure 10–12. A process fan is to run only when all of the following conditions are met:

1. Input 1 is off.
2. Input 2 is on and input 3 is on, or both 2 and 3 are on.
3. Inputs 5 and 6 are both on.
4. One or more of inputs 7, 8, or 9 is on.

The three B examples involve converting PC ladder diagrams to gate diagrams. Conversion to Boolean is an added option. B1, shown in figure 10–13, is a fundamental conversion. Series contacts are converted to AND gates. Parallel contacts are converted to OR gates. Then, the combinations are treated in the same manner—series combinations to AND and parallel combinations to OR.

Figure 10–14 illustrates example B2. B2 is more involved than B1 and requires more gates for the gate diagram. It also requires an input inversion. Since contact 10 is normally closed in the PC ladder diagram, its state must be inverted. The inversion is accomplished at the input of logic gate OR 22. By convention, the dot performs the same function as a NOT gate.

Figure 10-10
Example A2:
Word Description
Conversion

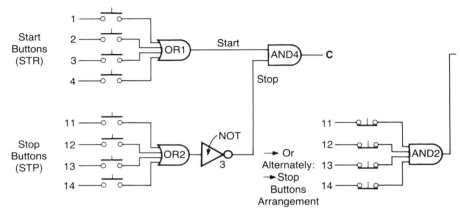

Gate Logic

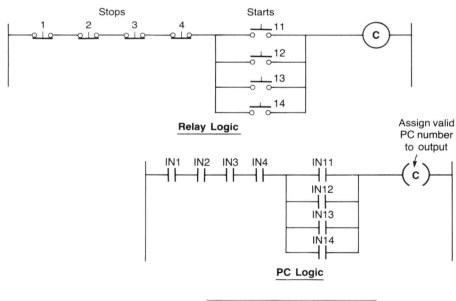

Relay Logic

PC Logic

$$(STR1 + STR2 + STR3 + STR4) \cdot \overline{(STP1 + STP2 + STP3 + STP4)} = C$$

or

$$(STR1 + STR2 + STR3 + STR4) \cdot \overline{(STP1 + STP2 + STP3 + STP4)} = C$$

Boolean (Ref.)

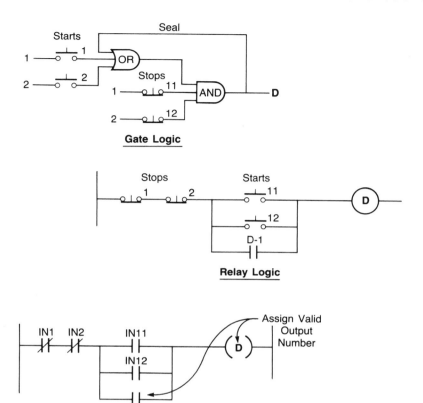

Figure 10–11
Example A3:
Word Description
Conversion

$$(\mathrm{STR1} + \mathrm{STR2} + \mathrm{D}) \cdot \overline{\mathrm{STP1}} \cdot \overline{\mathrm{STP2}} = \mathrm{D}$$
Boolean (Ref.)

Example B3 is a more advanced circuit requiring many gates and two inversions. The construction of the equivalent gate diagram follows the principles of the previous two examples. It is shown in figure 10–15.

The three C examples convert given digital gate diagrams into ladder diagrams. The ladder diagrams are drawn for PC logic only. As in the A and B examples, the Boolean equivalent is given for reference.

Example C1 in figure 10–16 is a fundamental gate-to-ladder-diagram conversion. Two different AND gates feed one OR gate. If either or both of the feeder gates are on, the OR gate is on and output R will be on.

Example C2 has two OR gates and one AND gate all feeding an AND gate. The OR gate is converted to parallel contacts in the ladder diagram. The AND gate is converted to series contacts or to series groups of con-

Figure 10-12
Example A4:
Word Description
Conversion

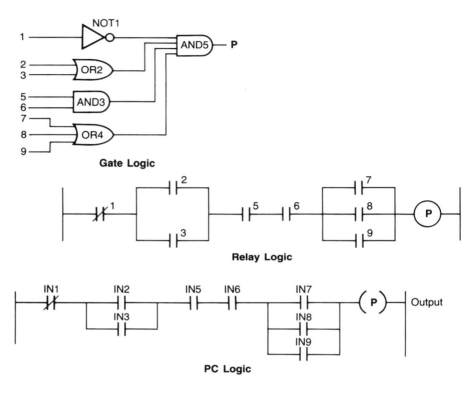

Gate Logic

Relay Logic

PC Logic

$$(\overline{1}) \cdot (2 + 3) \cdot (5 \cdot 6) \cdot (7 + 8 + 9) = E$$

Boolean (Ref.)

Figure 10-13
Example B1:
Ladder-Diagram-
to-Gate Con-
version

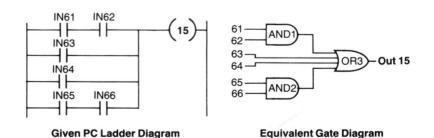

Given PC Ladder Diagram   **Equivalent Gate Diagram**

$$(61 \cdot 62) + (63) + (64) + (65 \cdot 66) = 15$$

Equivalent Boolean Expression (Ref.)

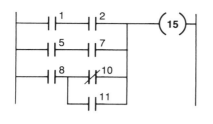

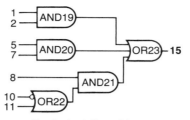

Figure 10-14
Example B2:
Ladder-Diagram-
to-Gate Con-
version

Given PC Ladder Diagram   **Equivalent Gate Diagram**

$$(1 \cdot 2) + (5 \cdot 7) + [(8) \cdot (\overline{10} + 11)] = 15$$

Equivalent Boolean Expression (Ref.)

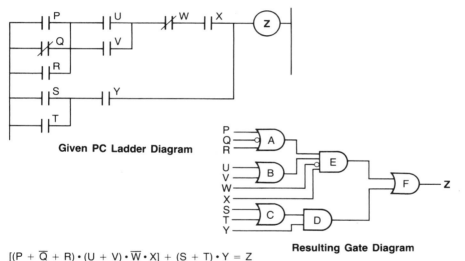

Given PC Ladder Diagram

Figure 10-15
Example B3:
Ladder-Diagram-
to-Gate Con-
version

Resulting Gate Diagram

$$[(P + \overline{Q} + R) \cdot (U + V) \cdot \overline{W} \cdot X] + (S + T) \cdot Y = Z$$

Equivalent Boolean Expression (Ref.)

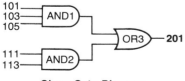

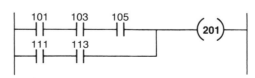

Figure 10-16
Example C1:
Gate-To-Ladder
Conversion

Given Gate Diagram   **Resulting PC Ladder Diagram**

$$(101 \cdot 103 \cdot 105) + (111 \cdot 113) = 201$$

Equivalent Boolean Expression (Ref.)

101

tacts. The only new concept introduced is the dot on the H input. The dot means the input is inverted. It is the same as if a NOT gate were in the line between the H and the input point of AND gate 3. Figure 10–17 shows the conversion.

Example C3 in figure 10–18 is another gate-to-ladder-diagram conversion. A new feature introduced is an input being fed to two different places, which is shown in the resulting ladder diagram. Two different contacts are needed in the ladder diagram for it to be equivalent to the one gate input number.

Figure 10–17
Example C2:
Gate-To-Ladder
Conversion

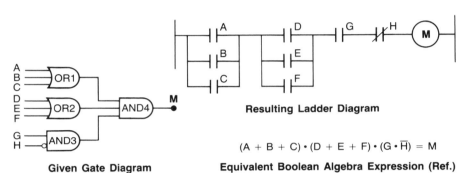

Given Gate Diagram

Resulting Ladder Diagram

$$(A + B + C) \cdot (D + E + F) \cdot (G \cdot \overline{H}) = M$$

Equivalent Boolean Algebra Expression (Ref.)

Figure 10–18
Example C3:
Gate-To-Ladder
Conversion

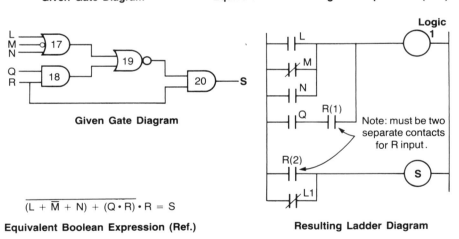

Given Gate Diagram

Resulting Ladder Diagram

$$\overline{(L + \overline{M} + N)} + (Q \cdot R) \cdot R = S$$

Equivalent Boolean Expression (Ref.)

EXERCISES

For exercises 1 through 4, convert the word description to:

    A - Gate symbols
    B - PC/Relay logic ladder diagram

1. Switch 8 and switch 11, plus either switch 22 or switch 34, must be on for output 67 to be on.

2. For output 7 to be on, input 6 must be off and either input 8 or input 9 must be on. In addition, one of inputs 1, 2, or 3 must be on.

3. For output H to be on, input A must be on and both inputs C and D must be off. In addition, one or more of inputs E, F, and G must be off.

4. There are four push-button stations controlling a fan. Each fan has a start and stop button. Two door interlocks must be closed before the fan may run. Pushing any start button will make the fan run, and the fan is sealed on when the start button is released. Pushing any stop button turns the fan off and also prevents the fan from starting or running.

For exercises 5 through 7, convert the gate diagrams given in figures 10–19, 10–20, and 10–21 to PC ladder diagrams.

For exercises 8 through 10, convert the PC ladder diagrams given in figures 10–22, 10–23 and 10–24 to gate diagrams.

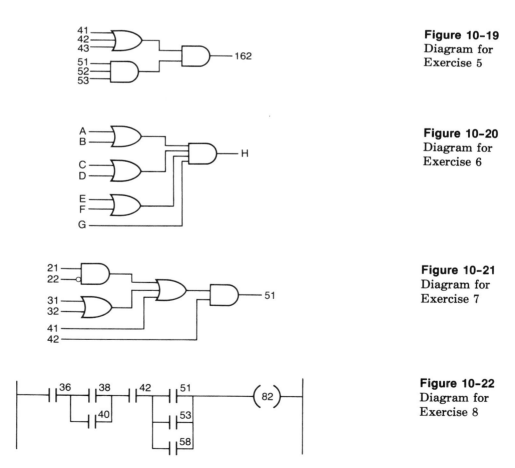

Figure 10–19
Diagram for
Exercise 5

Figure 10–20
Diagram for
Exercise 6

Figure 10–21
Diagram for
Exercise 7

Figure 10–22
Diagram for
Exercise 8

Figure 10-23
Diagram for
Exercise 9

Figure 10-24
Diagram for
Exercise 10

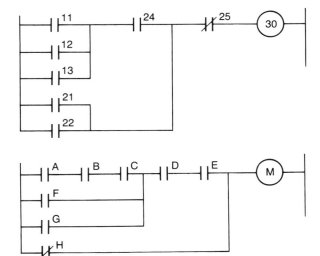

11. (Optional) Convert the diagrams of exercises 1 through 10 to Boolean algebra expressions.

SECTION THREE

BASIC PC FUNCTIONS

Programming On-Off Inputs to Produce On-Off Outputs

11

At the end of this chapter, you will be able to

☐ Describe the contact (input) function of the PC.
☐ Describe the coil (output) function of the PC.
☐ Convert industrial control problems to PC logic diagrams.
☐ Show the advantage of PCs over relay logic in simpler connection diagrams and wiring.
☐ Develop a PC ladder circuit for an industrial problem.

INTRODUCTION

Chapter 11 will show how to program a PC for circuit operations. The chapter will deal only with straight inputs and outputs. Subsequent chapters will discuss counters, timers, and other advanced functions.

In PCs, the internal symbol for any input is a contact. Each input terminal on the input module has a number. Internally, the PC represents the input status of each input by a contact status of the same number. Similarly, in most cases, the internal PC symbol for all outputs is a coil. Other functional blocks can represent an output and will be covered in subsequent chapters. Each internal coil has a number. When on, the corresponding output of the same number will be on at the output module.

Chapter 11 begins with basic, simple control circuits and then progresses to more complicated control circuits. The circuits are compared to relay logic for ease of understanding. You will see that PC circuit connecting remains relatively easy as circuits become more complicated. For relay-based control circuits, connection wiring increases in complexity as the circuit complexity increases. A somewhat complicated industrial-type problem is also included in the chapter.

There are some unique contact/coil functions in the PC not easily available with relays. These functions include the latch/unlatch coil, which is covered in this chapter. Other special coil functions, such as transitional contacts, are discussed in chapter 24.

INPUTS/CONTACTS

Contacts in a PC system are mainly concerned with inputs. Each input to an input module has a corresponding programmed PC contact in the CPU. However, all contacts in an internal program do not need to have a corresponding input.

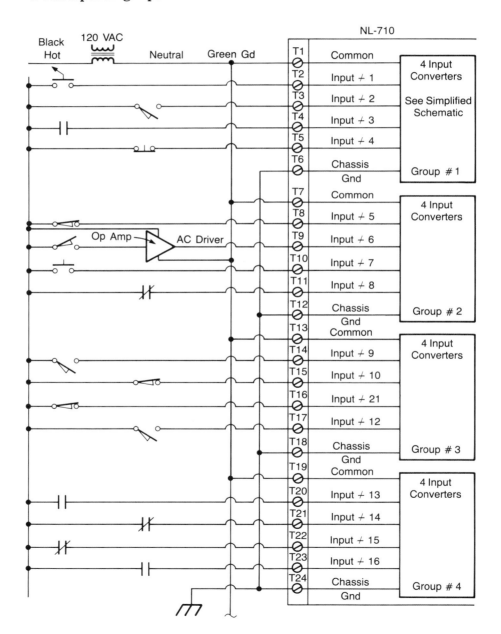

Figure 11-1
Typical PC
Input Scheme

In a PC system, each input is assigned a number on the input module and in the CPU. The number may be a reserved block of numbers or letters. In other PCs, some prefix is used, such as IN. In a prefix system, the fifth input would correspond to the PC program number IN 0005. A typical input scheme is shown in figure 11-1. The input terminals correspond to a series of numbers, such as IN 0001 through IN 0016. The numbers of a module are set by DIP switches, as described in chapters 2 and 3.

Suppose that we apply power to terminal 5. All contacts programmed in the PC as IN 0005 will change states. The input is examined and the proper action takes place. All normally open IN 0005 contacts will go to the closed state in the PC program. Also, all normally closed IN 0005 contacts go to an open state. This opening of normally closed contacts is a key concept in understanding PC programming.

Some typical input contact devices are shown in figure 11-2. More are shown in appendix C. Note that the value of input supply voltage must correspond to the voltage rating of the input module.

There is one other key point on contacts as related to inputs. Suppose a contact in the internal program is labeled IN 0018. Also suppose that the only inputs connected are IN 0001 through IN 0016. Would the IN 0018 programmed contact ever change state from external signals? No.

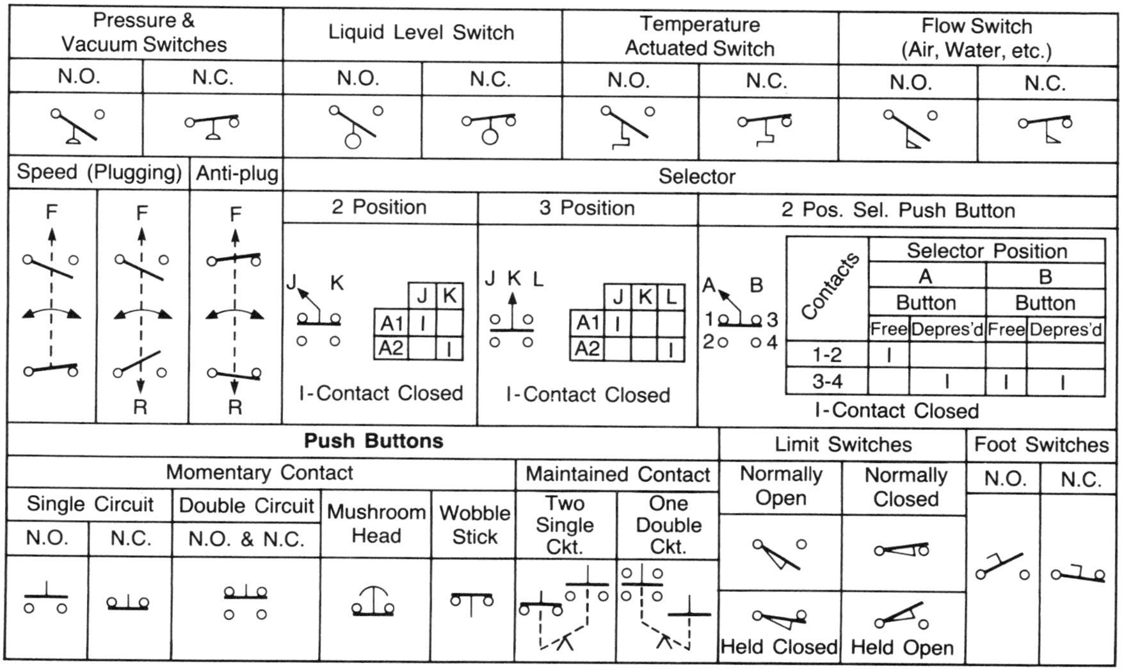

Figure 11-2
Typical PC Input Devices

There is no energizing signal available from an input module to have an effect on the internal CPU status.

OUTPUTS/COILS

Coils in an internal PC program are related to output signals that are sent to external devices. An output is energized through the output module when its corresponding coil number is turned on in the PC ladder diagram. Note that not all coils in a program have a corresponding output. Many

Figure 11-3
Typical PC
Output Scheme

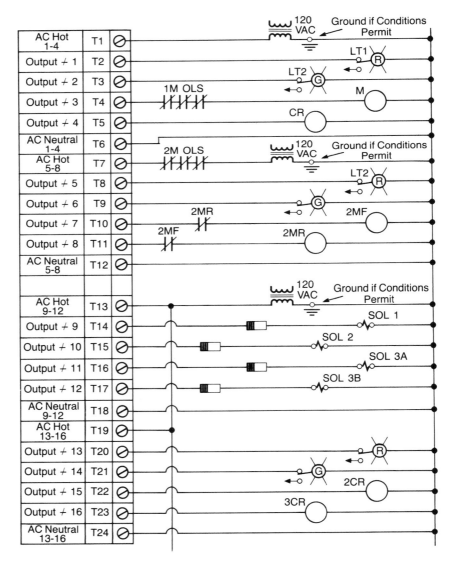

coils are used for internal logic only. A typical output scheme is shown in figure 11-3. The output device's voltages and current requirements must be matched for the output module values.

In a manner similar to inputs, output numbers must correspond. For example, only outputs CR 0017 through CR 0032 are connected to the CPU through an output module. If program coils have numbers such as CR 0014 and CR 0034, neither will affect any output. There is no corresponding coil for the output signal to affect. If CR 0018 is turned on, output 18 will turn on.

Some typical output devices for coil outputs are shown in figure 11-4. A key point previously mentioned is the presence of a small output module leakage current when the PC output is off. The leakage current must be considered if the output device is sensitive to a low value of voltage. The output device might not turn off even when the output module is technically in the off state. Figure 11-4 illustrates some of the typical output devices used in processes. More are shown in appendix C.

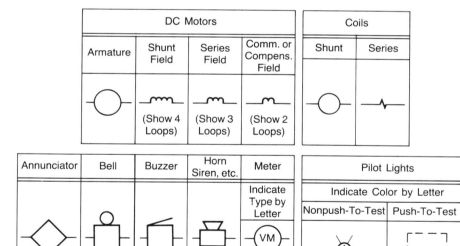

Figure 11-4
Typical PC
Output Devices

CONTACT AND COIL PROGRAMMING EXAMPLES

This section will illustrate seven representative examples of PC programming using contacts and coils. The first six examples range from basic to intermediate. The seventh is a more complex industrial problem. For the first three examples, both PC and relay logic solutions will be shown.

For examples three and beyond, only the PC connection diagram will be given in the problem solution.

The seven examples are:

1. Simple one-contact, one-coil circuit.
2. Standard start-stop-seal circuit. Alternate latch-unlatch circuit.
3. Forward-reverse-stop with mutual interlocks.
4. Forward-reverse-stop with direct reversal.
5. Start-stop-jog.
6. Alarm system.
7. Drill press industrial control.

The first example, A, is a simple circuit with one switch as a contact and one output as a coil. As the switch is opened or closed, the output goes on or off. Figure 11–5 shows the ladder diagrams for relay logic and ladder logic. They are identical for this example.

Example B is a start-stop-seal circuit. When the start button is depressed, the coil energizes. When start is released, the coil remains on. It is held on by a sealing contact that is in parallel with the start button.

Figure 11-5
Example A:
Simple One-
Switch, One-Coil
Control

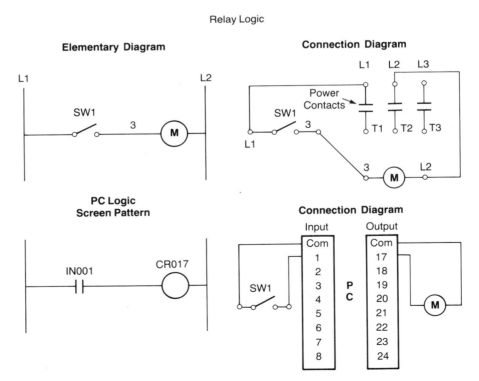

The seal contact closes when the output coil goes on. If the stop button is depressed, the coil goes off and stays off. Also, if the control power goes off, the coil goes off. The advantage of example B over A is that when failed control power returns, start must be depressed to reenergize the coil. For example A, the coil would immediately restart, possibly posing a safety hazard to an unsuspecting operator or repair person.

The relay logic and PC logic diagrams for example B are shown in figure 11-6. A major difference between relay and PC connections is in the physical location of the seal contact. In relay logic, the seal contact is attached physically to, and goes on and off with, the output coil. In the PC

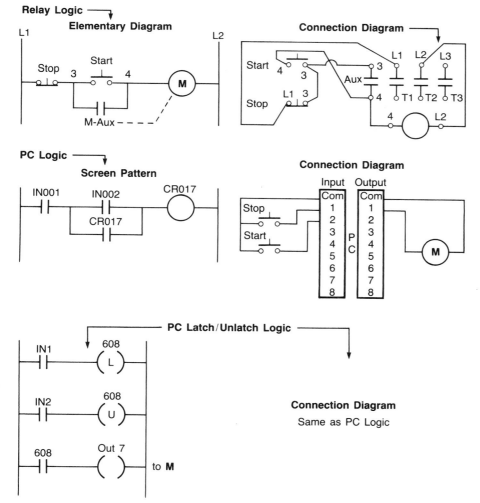

Figure 11-6
Example B:
Standard Start-
Stop-Seal Circuit

control, the seal is generated internally in the PC logic. In PC logic, the seal closes or opens as the output coil goes on or off.

Figure 11–6 also illustrates the latch-unlatch function found in many types of PCs. One input switch latches the output on. A different input switch latches the output off. The latch-unlatch PC coil may be an internal one. A contact from its coil may have to be used to control a coil that corresponds to an output terminal.

Note that in figure 11–6, a PC input, IN 0002, must be energized to turn the coil off. The safety implications of this situation were discussed in chapter 7. A safety-related circuit would not use a latch-unlatch relay.

The third example, C, is a standard forward-reverse circuit. Each direction's coil has its own start button. The single stop button stops either coil's operation. In this circuit, you must push stop before changing direction. Interlocks are provided so that both outputs cannot be energized at the same time. This particular circuit works for other control applications, as well. It could be used for low-speed/high-speed or part-up/part-down control systems. Figure 11–7 illustrates example C's circuit. IN 0001 stops operation in either direction. IN 0002 is for forward, CR 0017, and IN 0003 is for reverse, CR 0018.

At this point, we begin to see that the PC connections are simpler than connections for relay logic. Compare the two connection diagrams: the relay logic connections are quite complicated in comparison to the PC connections.

Example D, which is shown in figure 11–8, is similar to example C. The major difference is that in D you may go directly from one direction to another without first depressing stop. Another difference is the added directional pilot light indicators. The key for input identification is shown on the diagram.

The quick reversal may be desirable in some operational applications; however, in some cases it may not be a good system. If, for example, we have a large flywheel as an output device connected to an electric motor, applying instant reversal would cause undue stress on the motor, the power distribution system, and the mechanical parts and mountings. A more advanced circuit with a time delay would be required.

Next is example E. In some cases, you may wish to have the output on momentarily only at times. The momentary on is called *jog*. At other times you might like the output on all the time, as in previous examples. Two possible circuits for start-stop-jog are shown in figure 11–9. The PC program and the PC connection diagram are included also. Note that it is necessary to push stop before going from run to jog in the circuit illustrated here.

Example F is an alarm system. There are four hazard inputs to the alarm system which go on as some operational malfunction occurs. We

Relay Logic

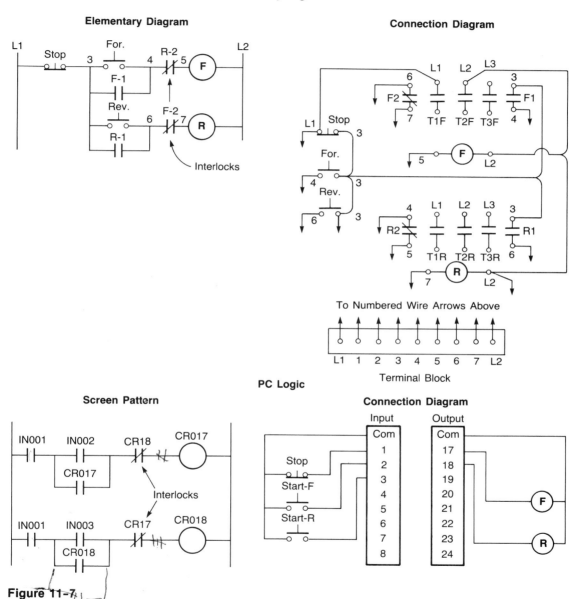

Elementary Diagram

Connection Diagram

To Numbered Wire Arrows Above

Terminal Block

PC Logic

Screen Pattern

Connection Diagram

Interlocks

Figure 11-7
Example C: Forward-Reverse Control

Relay Logic (Reference)

3 Wire Control - Reversing Starter
with Pilot Lights to Indicate
Direction Motor is Running

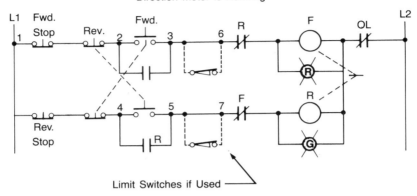

Limit Switches if Used

PC Logic

See Note

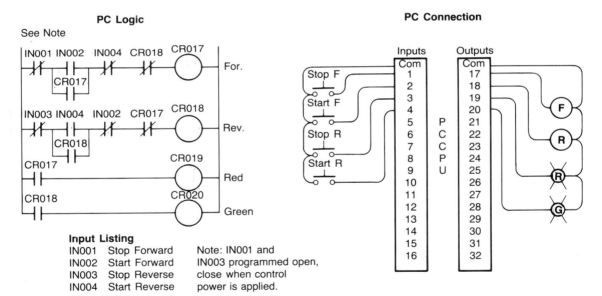

Input Listing

IN001	Stop Forward
IN002	Start Forward
IN003	Stop Reverse
IN004	Start Reverse

Note: IN001 and
IN003 programmed open,
close when control
power is applied.

PC Connection

Figure 11-8
Example D: Instant Forward-Reverse Change Circuit

116

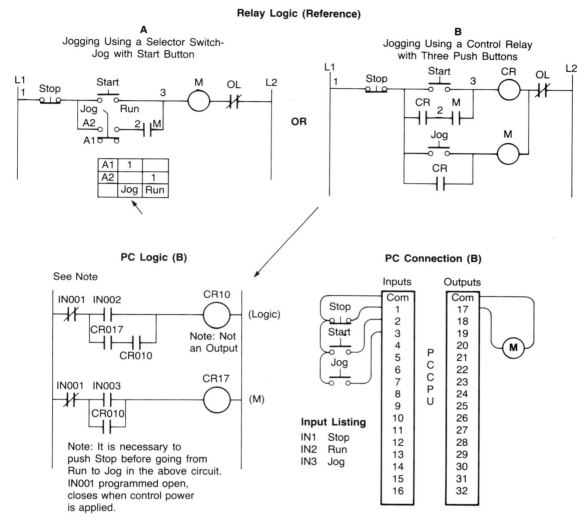

Relay Logic (Reference)

Figure 11-9
Example E: Start-Stop-Jog

will not define what the hazards are; for PC operation illustration, we will only use the fact that there are four. The system operates as follows:

☐ If one input is on, nothing happens.
☐ If any two inputs are on, a red pilot light goes on.
☐ If any three inputs are on, an alarm siren sounds.
☐ If all four are on, the fire department is notified.

Since this example is more involved than the previous ones, we will take time to specify input and output numbers. The PC program numbers for the inputs and outputs are assigned as follows:

	Inputs	Outputs	
A	IN 0001	Red Pilot Light	CR 0017
B	IN 0002	Alarm (Siren)	CR 0018
C	IN 0003	Fire Department Notify	CR 0019
D	IN 0004		

A PC logic diagram to accomplish the circuit requirements is shown in figure 11–10. One final note for example F: connecting the PC terminal to the output alarms is very simple; if you had a relay-switch system, the connections would be very involved and complicated.

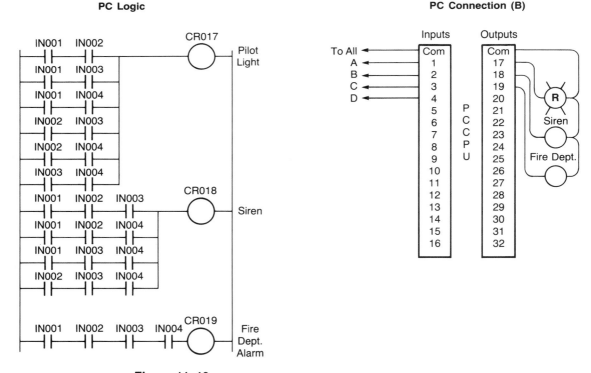

Figure 11–10
Example F: Alarm System

INDUSTRIAL PROCESS EXAMPLE

Example G is a more involved problem than the previous ones in this chapter. To formulate a control system, we will generally follow the procedural steps of chapter 7. The problem involves a semi-automatic drill press operation as shown in figure 11–11.

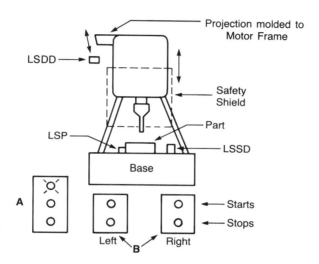

Figure 11-11
Drill Press
Operation Layout

The initial position of the drill press spindle is at the top, as shown. A part to be drilled is placed under the spindle. The drill is then to come down after two start buttons are depressed. (Two push buttons are recommended to assure that both hands are out of the way.) The drill spindle rotates as it is brought downward. Downward spindle force is furnished by a pneumatic air cylinder pushing against an upward return spring; pneumatic control air is supplied through an electrical solenoid. When the spindle is completely down and the drill bit goes through the part to be drilled, a down sensor is actuated. The solenoid is then deenergized and the drill returns up by means of the return-up spring. When the spindle is completely back up, the system is to be reset to the off condition. If no part is in place initially, the drill spindle cannot descend.

In addition to the operation previously described, a safety shield will be included. For extra safety, a screen shield comes down before the drill can start down. The shield returns up at the same time as the drill by its own spring return. The shield's descent is powered by its own separate pneumatic solenoid.

When the stop button is pushed at any time, the drill and shield return up. Note that this could be a safety hazard. More circuitry would be needed to stop the spindle where it is when the stop button is depressed.

There are a number of procedural steps to go through to arrive at a solution. Previous chapter examples have not been complicated, and we performed their procedural steps informally. The steps recommended for a problem of this type are:

1. Define the process operation and list the step-by-step sequence of operation.
2. Define and list the input and output devices and sensors required for proper operation.
3. Assign corresponding PC numbers to the input and output devices.
4. Draw up the PC scheme. Note that margin notes are helpful.
5. Enter the program into the PC.
6. Optional step: check the program sequence by using the FORCE mode. (The FORCE mode will be explained in detail in chapter 14.)
7. Wire the PC system to a simulator and check its operation.
8. Check the actual process operation. Try various out-of-sequence operations to check for hidden safety defects or sequencing problems. For example, what happens if the power fails when the spindle is half-way down.
9. Make modifications as required.

Step one is to list the sequence.

1. Push system start.
2. Put part in place to actuate LSPP.
3. Push the two start buttons simultaneously.
4. Safety shield comes down, actuating LSSD.
5. Drill starts rotating and descends.
6. Drill at bottom actuates LSDD.
7. System shuts down. Drill and shield return up by springs.
8. System is reset.

Note that pressing stop at any time stops the sequence and resets the spindle to the top.

Step two is to list the input and output devices.

☐ system start switch
☐ system stop switch—stops everything
☐ system pilot light
☐ shield and drill start—left hand switch
☐ shield and drill start—right hand switch
☐ position indicator—part in place

☐ position indicator—shield down
☐ position indicator—drill down

Step three is to assign input and output numbers to all components. This includes switches and sensors.

Input	Output
IN 0001 System start	OUT 0017 System pilot light
IN 0002 System stop	OUT 0018 Shield down solenoid
IN 0003 LSPP—part in place	OUT 0019 Motor rotate motor
IN 0004 Left start	OUT 0020 Drill down solenoid
IN 0005 Right start	
IN 0006 Left stop	
IN 0007 Right stop	
IN 0008 LSSD—Shield down	
IN 0009 LSDD—Drill down	

Step four is to sketch the PC system.
Step five is to load the sketch into the CPU. The ladder diagram formulated is shown in figure 11–12.

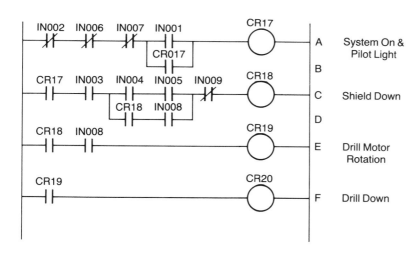

Figure 11-12
Drill Press PC
Control Circuit

A sequence by line for the ladder diagram in figure 11–12 is:

☐ Line A. Push start, IN 0001. CR17 goes on.
☐ Line B. CR 0017 seals on.

☐ Line C. Put part in place. IN 0003 closes. IN 0009 was already closed. Push right and left starts, IN 0004 and 0005. CR 0018 on. Shield starts down.

☐ Line D. Start buttons IN 0004 and 0005 must still be pushed. CR 0018 is closed on. When shield reaches down position, LSSD closes, IN 0008. CR 0018 seals on.

☐ Line E. CR 0018 and 0019 are closed. CR 0019 on, motor rotates.

☐ Line F. CR 0019 starts motor down, CR 0020.

☐ Line G. Drill reaches lower position and hits LSDD, IN 0009. CR 0018, 19, and 20 all go off. Motor goes off and shield and drill spring-return up.

Step six is an optional FORCE analysis.

Step seven is to wire the system to a simulator. A wiring scheme appears in figure 11–13. Note the connection diagram's simplicity for the PC—only five wires.

Step eight, circuit operation, and step nine, modifications, would follow after an analysis of the drill press's actual operation.

Figure 11-13
Input and Output
Module Wiring
For Drill Press

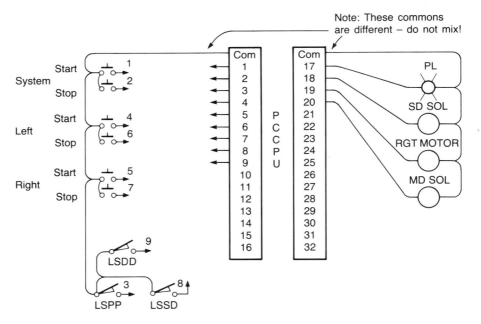

EXERCISES

Construct PC ladder diagrams for the listed problems. A sequence could be written first, if necessary. As an option, show the input and output modules along with the device-to-terminal electrical connections.

In the laboratory, you may load the PC CPU with your program. Connect the PC to a simulator. Check out the proper operation of the circuit by running the program sequence.

1. A fan is to be started and stopped from any one of three locations. Each location has a start and stop button. Refer to example B in the chapter. Note that normally closed stops should be in series and normally open starts in parallel.

2. A two-way hydraulic cylinder has two solenoids controlling it. Energizing one solenoid causes the cylinder to extend and energizing the other solenoid causes it to retract. A limit switch at each end indicates full retraction or full extension. Use two start/stop, three-wire controls, one for each direction. Construct a two-directional control system, including interlocks, to control the solenoid. Refer to example C in the chapter.

3. A milling machine (M) and its lubrication pump (L) both have three-wire, start/stop control systems. L must be running before M can be started. Furthermore, if L stops, M must also stop.

4. Two separate start-stop-jog control stations are required for a pump motor. Refer to example E in the chapter.

5. There are three machines, each with its own start/stop buttons. Only one may run at a time. Construct a circuit with appropriate interlocking.

6. Repeat exercise 5, except that any two may run at one time. Also, any one may run by itself.

7. A temperature control system consists of four thermostats. The system operates three heating units. Thermostats are set at 55, 60, 65, and 70 °F. Below 55 °F, three heaters are to be on. A temperature between 55 to 60 °F causes two heaters to be on. For 60 to 65 °F, one heater is to be on. Above 70 °F, there is a safety shutoff for all three heaters in case one stays on by mistake. A master switch turns the system on and off.

8. Create a PC system in a manner similar to example G in the chapter for the problem in figure 11–14.

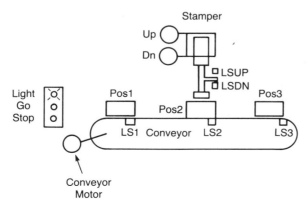

Figure 11–14
Diagram for
Exercise 8

When a part is placed on the conveyor at position 1, it automatically moves to position 2. Upon reaching position 2, it stops and is stamped. After stamping, it automatically moves to position 3. It stops at 3, where the part is removed manually from the conveyor.

Assume only one part is on the conveyor at a time. Add limit switches, interlocks, push buttons, etc., as required.

If you become stuck at the middle station, you may add a manual restart switch for this point on the conveyor.

Timers

12

INTRODUCTION

The most commonly used process control device after coils and contacts is the timer. The most common timing function is the time delay on, which is the basic function. There are also many other timing configurations, all of which can be derived from one or more of the basic time delay on functions. PCs have the one basic function timer capability in multiples. This chapter will illustrate the basic PC time delay on function, and seven other derived timing functions. Typical of the derived functions are: time delay off, interval pulse timing, and multiple pulse timing of more than one process operation.

There is normally only one of two types of the basic PC timing functional blocks in a PC. The timing block functions are used with various contact arrangements and in mutiples to accomplish various timing tasks. Typical industrial timing tasks include timing of the intervals for welding, painting, and heat treating. Timers can also predetermine the interval between two operations. With a PC you can utilize as many timer blocks as you need, within the PC memory limitations.

What does the PC timer function replace? Detailed descriptions of the traditional industrial timers may be found in controls texts, including the controls texts listed in the bibliography. Symbols for conventional timers may be found in appendix C. PC timer functions can replace any of these industrial timers. Whether the industrial timer is motor driven, RC time constant, or dash-pot, it can be easily simulated by a PC.

The digital, solid-state, electronic timer is one technological step above the three types of industrial timers just listed. These digital timing devices are also discussed in various controls texts. The PC timing function is more versatile and flexible than either the industrial or the digital electronic timers.

One major advantage of the PC timer is that its time may be a programmable variable time as well as a fixed time. The variable time interval may be in accordance with a changing register value. Another advantage of the PC timer is that its timer accuracy and repeatability are extremely high, since it is based on solid state technology.

THE BASIC PC TIMER FUNCTION

A single input timer is sometimes used in PCs. An example is shown in figure 12-1. Energizing IN 0001 causes the timer to run for 4 seconds. At the end of 4 seconds the output goes on. Then when the input is deenergized, the output goes off and the timer resets to 9. If the input IN 0001 is turned off during the timing interval, the timer never comes on and the timer resets to 0.

There are some operational disadvantages of the single-input type that are overcome by the double-input PC timer. The double-input type is more commonly used in PCs. We shall use the double–input timer throughout this chapter. Figure 12-2 shows the layout of the double–input timer for two typical formats.

The lower input in figure 12-2 is the enable/reset line, which allows the timer to run when energized. When deenergized, the timer is kept at 0, or reset to 0. The upper line causes the timer to run when the timer is enabled. When enabled, the timer runs as long as the run input is energized. If run is deenergized while the timer is running, the timing stops where it is and does not reset to 0.

Figure 12–1
Single-Input
Timer

Suppose IN 0002 is closed and IN 0001 is turned on. After 6 seconds, IN 0001 is opened. The timer retains a count of 6. Timing has not reached the preset value of 14 seconds and the timer output is still off. The timer does not reset unless IN 0002 is opened. Suppose that sometime later IN 0001 is reclosed. After 8 more seconds of IN 0001 being closed, the timer coil will energize, since $6 + 8 = 14$.

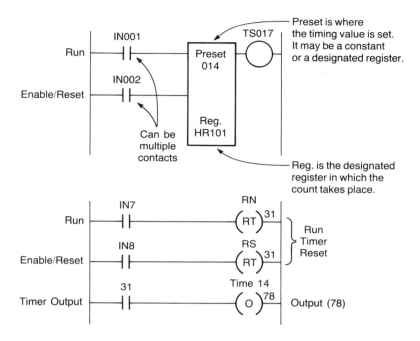

Figure 12-2
Double-Input
Timer

Defining Timer Status With X and O

An added help in defining timer contact status is shown in figure 12-3. There are three states in a timing cycle: (a) initial or reset state, (b) state during timing, and (c) state after timing is complete. A system of X for on and O for off is normally used. The chapter examples illustrate the use of this convention.

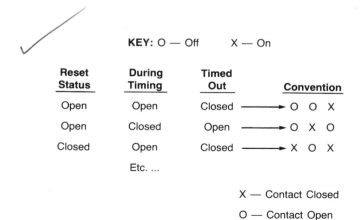

Figure 12-3
Sequencing Chart

EXAMPLES OF MAJOR TYPES OF TIMER FUNCTIONS

Some commonly used timer functions that will be illustrated are

- ☐ On delay. Output B comes on at a specific set time after output A is turned on. When A is turned off, B also goes off.
- ☐ Off delay. Both A and B have been turned on at the same time. Both are in operation. When A is turned off, B remains on for a specific set time period before going off.
- ☐ Limited ON time. A and B go on at the same time. B goes off after a specific set time period, but A remains on.
- ☐ Repeat cycling. An output pulses on and quickly off at a constant preset time interval.
- ☐ "One shot" operation. Output B goes on for a specified time after output A is turned on. Output B will run for its specified time interval even if A is turned off during the B timing interval.
- ☐ Alternate on and off of two outputs. An example of this timing application is two alternately flashing signal lights. The time on for each of the two lights may be the same, or the two times could be set to different intervals.
- ☐ Multiple on delay or multiple off delay. Two or more different events are timed from the same initial time reference point.
- ☐ Interval time within a cycle. We may require that an output is to come on 7.5 seconds after system start up, remain on for 4.5 seconds, and then go off and stay off. The interval would then be repeated only after the system is shut off and then turned back on.

There are other timing examples we could illustrate as well; however, these eight examples are representative of PC timing capabilities.

EXAMPLES OF TIMER FUNCTIONS

We will give examples of each of the eight listed timing systems. In most cases, each example is illustrated in terms of an industrial problem. Each example includes a diagram showing time versus output on-off status. The time-status diagrams include the X and O designations for reference.

Example A is the simplest form of time delay. When the circuit is turned on, one action takes place. Then, at a specified time later, another action occurs. Both relay logic and PC logic diagrams are shown in this example only for comparison. Subsequent examples will have PC diagrams only. Figure 12–4 shows the program for example A.

The sequence for example A is:

1. When switch 1 is turned on, light A lights.
2. Eight seconds after A lights, B lights.
3. Both lights go off or stay off whenever switch 1 is opened.

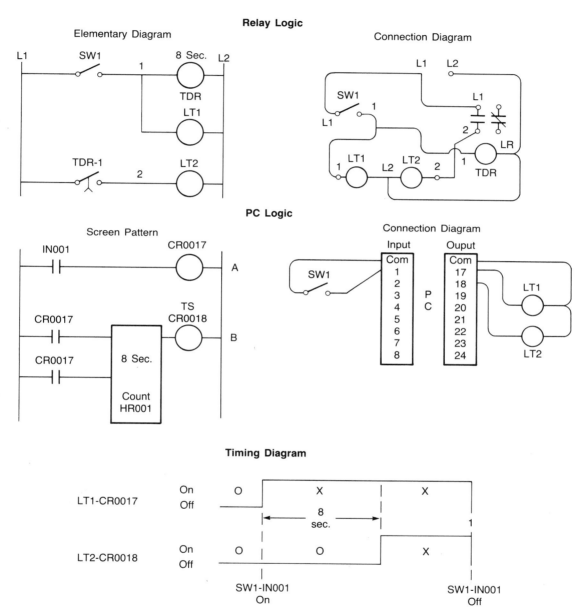

Figure 12-4
Example A. Time Delay On

Example B is an off-delay circuit. A motor and its lubrication pump motor are both running. Lubrication for main motor bearings is required during motor coast-down. After the main motor is shut off, the lubricating pump remains on for a time corresponding to coast-down time. In this example, a lubricating pump remains on for 20 seconds after the main system is shut down. Figure 12–5 shows the required program.

Example C1 illustrates a situation in which two inputs go on at the same time. Then, one of them is to go off after a preset period of time. Figure 12–6 shows this situation for two outputs. One output, A, stays on; the other output, B, turns off at the end of the timing interval. Resetting is accomplished by turning IN 0001 and 0002 off.

Figure 12–7, example C2, shows a multiple application timing system. Three outputs turn on at the same time. One stays on. Another, M, shuts off after 8 seconds. The third output, N, shuts off after 14 seconds.

Figure 12–5
Example B.
Time Delay Off

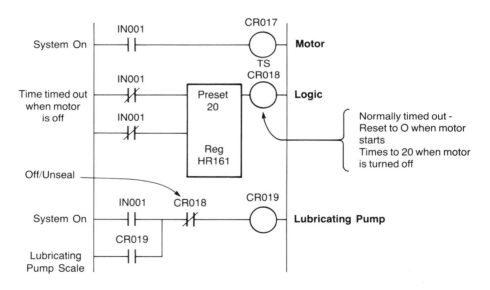

Timing Diagram

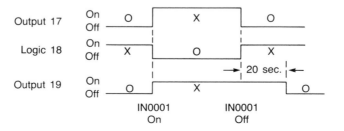

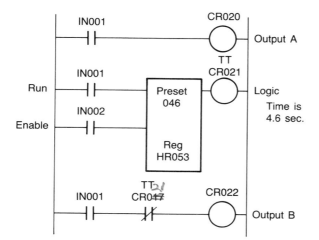

Figure 12-6
Example C1.
One Output,
Time Interval
On at Start

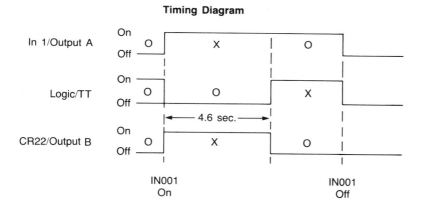

Example D is a pulsed timer. A short voltage pulse is produced every 12 seconds. The PC circuit shown in figure 12-8 will produce the required pulses. The timer is initially turned on by its "time" input. After the timing interval, the timer turns the output on. When the output goes on, one of its contacts, TT 0013, immediately opens and resets the timer to 0. When the timer is reset to 0, the output is turned off. Then, since the timer is also off, TT 0013 recloses, restarting the cycle. The pulsed on time is a very short, one-scan cycle time. The process repeats itself continuously.

Example E, shown in figure 12-9, is the "one shot" system. The output comes on after its specified time period even if the input is turned off during the timing period. IN 0011 must be opened and reclosed to reset the system.

Example F is an alternating two-output system. Figure 12-10 is an example of this application. It is a two-light, alternately flashing, PC pro-

Figure 12-7
Example C2.
Two Outputs,
Time Intervals
On at Start

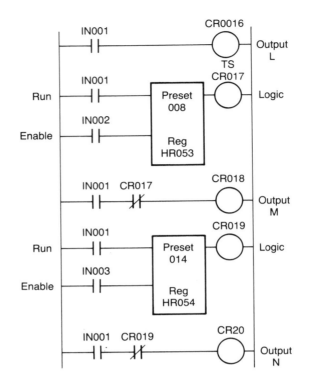

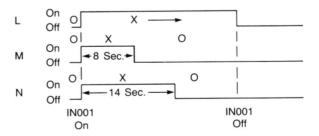

gram. Outputs 11 and 12 control the two lights. The outputs will alternate on-off and off-on every 5/10 of a second. The on times for each light can be varied by resetting the times in the functional block. The programmed times may be the same value, or each could be set at different time values. IN 0001 is the system on-off control.

Example G is for multiple timing of outputs and has two illustrations. Figure 12–11 is for dual time delay on. Figure 12–12 is for dual time delay off. Both figures are multiple applications of examples A and B, respectively.

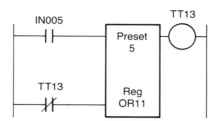

Figure 12-8
Example D.
Pulse Repetitive
Timer

Timing Diagram

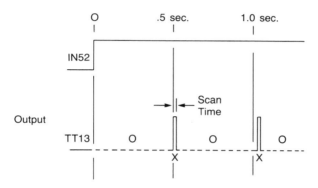

There are two ways to accomplish dual on delay. Both are shown in figure 12-11. The diagram on the left starts both timing intervals at the same time. The diagram on the right accomplishes the same timing, but chains the timers. On the right, timer 2 is started when timer 1 times out.

For dual time delay off, we will use an actual application for illustration. When the lights are turned off in a building, an exit door light is to remain on for an additional 45 seconds, and the parking lot lights are to remain on for an additional 3 minutes after the door light goes out. Figure 12-12 shows this dual PC off delay system.

Example H is for a timed interval of a number of seconds after the start of a process operation. This type of time interval is sometimes called an imbedded time interval. This operation uses the special operation of a fan. The fan is to come on 8.7 seconds after a system is turned on. It is then to run until 16 seconds after the system is turned on, which is a net time of 7.3 seconds. Figure 12-13 shows a PC program that accomplishes this time interval requirement.

AN INDUSTRIAL PROCESS TIMING PROBLEM

The following problem requires the use of multiple timing programming, as well as contact/coil logic. It consists of a single, operational, heat-

Figure 12-9
Example E.
"One Shot"
Timer Operation

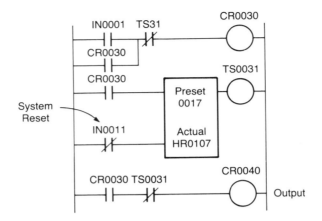

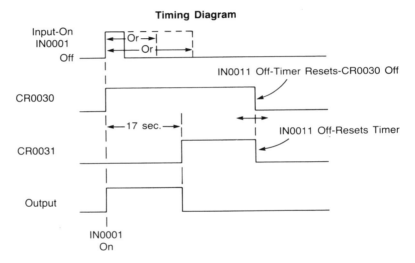

treating machine. The station carries out a surface-hardening process on a steel ring. Hardening is accomplished by heating the steel ring to a high temperature. Then it is immediately quenched (cooled very rapidly). The metallurgical result is a relatively hard surface on the steel ring.

The heating is done by a noncontact induction heating process. A high current in the circular coil around the outside of the part induces high circulating currents in the part. The part therefore heats up very rapidly. The coil has cooling water circulated through its outer half to keep the heating ring from overheating or even melting. The quench is then done by spraying cold water on the part. The quench water is pumped into the inner half of the induction coil. The spraying of the part with cold water through the many holes in the inside of the coil results in fast cooling, which produces a case-hardened surface.

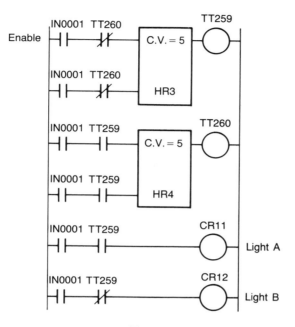

Figure 12-10
Example F.
Alternate
Flasher System

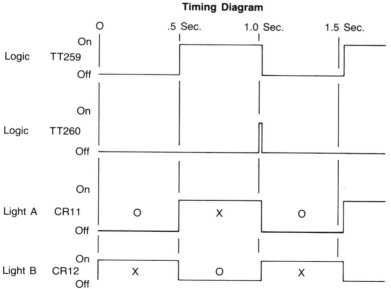

Timing Diagram

Figure 12-11
Example G1.
 Dual On Delay—
Two Schemes

Output 018 energizes after 7 sec. and 019 after 12 sec. (5 more) after IN001 is energized.

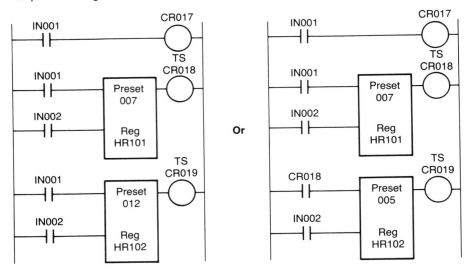

Or

Timing Diagram
IN002 is the enable/reset input.

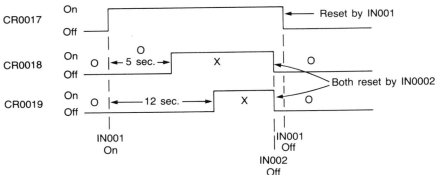

A mechanical layout of the process station is shown in figure 12–14. The processing sequence of this operation is:

1. Master push button is depressed, turning the system on.

2. Part is placed on the mandrel.

3. Both left and right start buttons are depressed.

4. At this point, or at any other time, pushing any stop button stops all processing action.

5. Part is raised from bottom to top by pneumatic air action. A solenoid valve supplies this air to a pneumatic elevating cylinder. The lower-

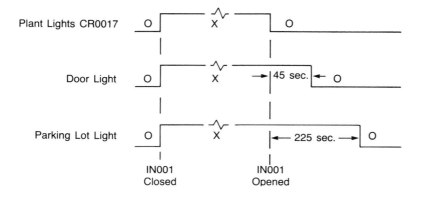

Figure 12-12
Example G2.
Dual Off Delay

IN001 —| |— CR017 — Lights - IN001 is closed during plant operation

TS
IN001 —|/|— Preset 045 — CR018 — Door Logic - Normally On (Timed Out)

IN001 —|/|— Reg HR075

IN001 —| |— CR18 —|/|— CR019 — Door Light

CR019 —| |—

TS
IN001 —|/|— Preset 225 — CR020 — Parking Lot Logic - Normally On (Timed Out)

IN001 —|/|— Reg HR076

IN001 —| |— CR20 —|/|— CR021 — Parking Lot Lights

CR021 —| |—

Timing Diagram

Plant Lights CR0017   O ___ X ___ O

Door Light   O ___ X → |45 sec.|← O

Parking Lot Light   O ___ X |← 225 sec. →| O

IN001 Closed   IN001 Opened

137

Figure 12-13
Example H.
Imbedded Inter-
val Timing

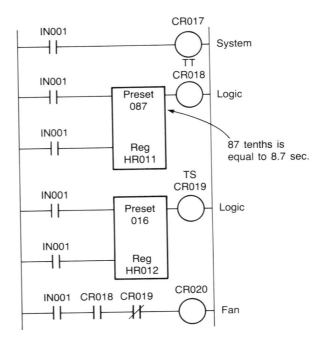

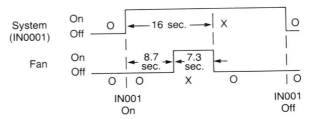

Timing Diagram

limit switch must be actuated before the part will rise. Lower-limit switch actuation indicates that there is a part on the mandrel and that the mandrel is down. Note that the limit switch opens as the part leaves the bottom position.

6. Mandrel makes contact with a limit switch at the top of the travel.

7. Heat comes on for 10 seconds and goes off.

8. Quench comes on for 8 seconds and goes off.

9. Part returns down by gravity and spring action. Upper limit switch becomes inactivated when the mandrel starts down.

10. Part and mandrel reach the bottom. Down limit switch is again actuated.

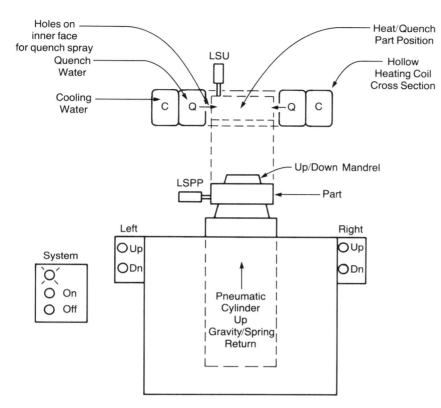

Figure 12–14
Heat/Quench
Station Layout

11. System should reset.
12. Part is removed.

Some optional features not included in this program are:

☐ If you assume the heat generator and both water pumps are on, interlocks could be added to insure they are running throughout the process.

☐ The same ring part could be processed two or more times. We could require the ring to be removed after step 12 before resetting takes place.

☐ Is proper temperature reached? A thermocouple sensor could be incorporated to monitor temperature.

☐ Manual controls for set-up could be added. These are Up, Heat, and Quench.

☐ Safety features could be added such as a safety shield that lowers during heat. Where does the process restart after interrupted power is restored?

☐ Other features as required.

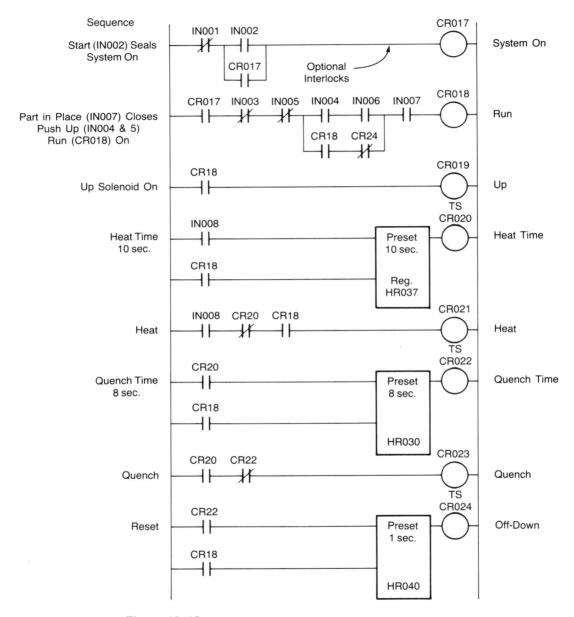

Figure 12-15
Heat/Quench Machine Program

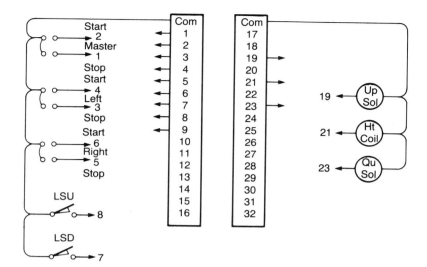

Figure 12-16
Connection
Diagram For PC
Module, Inputs,
and Outputs

The next step is to assign PC register or address numbers to the various inputs and outputs.

Inputs		Outputs	
0001	Master Stop	0019	Solenoid Valve—Up
0002	Master Start	0021	Heat On Contactor Coil
0003	Left Stop	0023	Quench Spray Water Solenoid
0004	Left Start		
0005	Right Stop		**Options**
0006	Right Start	0017	System On Pilot Light
0007	Limit Switch Down	0018	Machine On/Up Pilot Light
0008	Limit Switch Up		

A ladder diagram to carry out the process is then developed, as shown in figure 12–15.

The next step is to draw the connection diagram for the PC. There are eight input connections and three output connections (plus two for pilot lights, if included). The connection diagram is illustrated in figure 12–16.

The final two developmental steps are to program the PC for the process and make modifications as required.

EXERCISES

Write a PC program for these chapter problems, insert them into a PC, and test the programs for correct operation.

1. When a switch is turned on, C goes on immediately and D goes on 9 seconds later. Opening the switch turns both C and D off.

2. E and F are turned on by a switch. When the switch is turned off, E goes off immediately. F remains on for another 7 seconds and then goes off.

3. G and H both go on when an input is energized. G turns off after 4 seconds. H continues running until the input is deenergized. Turning input off at any time turns both outputs off.

4. Two pulsers start at the same time. Pulse output J is to pulse every 12 seconds. Pulse output K is to pulse every 4 seconds.

5. When L is turned on, M is to go on 11 seconds later. M goes on after 11 seconds, no matter how long L is turned on.

6. A. Two lights are to flash on and off at different intervals of 5 and 5 seconds, and 8 and 8 seconds. B. Two lights are to flash alternately, one for 5 seconds, one for 8 seconds.

7. There are four outputs: R, S, T, and U. R starts immediately when an input is energized. S starts 4 seconds later. T starts 5 seconds later than S. U goes on 1.9 seconds after S. One switch turns all outputs off.

8. Repeat exercise 7 for turning off delay on. S goes off 4 seconds after R. T goes off 6 seconds after R. U goes off 2.5 seconds after S.

9. An output pulse, V, is to go on 3.5 seconds after an input, W, is turned on. The V time-on interval is to last 7.5 seconds only. V is to go on again 3 seconds later for 5.3 seconds.

10. There are three mixing devices on a processing line: A, B, and C. After the process begins, mixer A is to start after 7 seconds elapse. Next, mixer B is to start 3.6 seconds after A. Mixer C is to start 5 seconds after B. All then remain on until a master enable switch is turned off.

11. When a start button is depressed, M goes on. Five seconds later, N goes on. When stop is pushed, both M and N go off. In addition, 6 seconds after M and N go off, fan F, which had previously been off, goes on. F remains on until the start button is again depressed, at which time it goes off.

12. A wood saw, W, a fan, F, and a lubrication pump, P, all go on when a start button is pushed. A stop button stops the saw only. The fan is to run an additional 5 seconds to blow the chips away. The lube pump is to run for 8 seconds after shutdown of W. Additionally, if the saw has run more than one minute, the fan should stay on indefinitely. The fan may then be turned off by pushing a separate fan reset button. If the saw has run less than one minute, the pump should go off when the pump is turned off. The 8-second time delay off does not take place for a running time of less than one minute.

Counters

13

At the end of this chapter, you will be able to

☐ Describe the PC counter function.
☐ List some of the major counting functions used in circuits and processes.
☐ Apply the PC counter function and associated circuitry to process control.
☐ Apply combinations of counters and timers to process control.

INTRODUCTION

PC counters have programming formats similar to timer formats. One input furnishes count pulses which the PC function analyzes. Another input carries out enable-reset, as it does in the timer function. Conventional counters replaced by this PC function can be mechanical, electrical, or electronic. Typical examples of solid-state counters may be found in manufacturers' manuals and controls texts.

Most PCs include both down counters and up counters, which function similarly. The up counter counts from 0 up to the preset count where some action takes place. The down counter goes from a preset number down to 0 where the action occurs. Having both up and down counters enables a common register to keep track of a net count. Use of two or more PC counters can be useful in controlling processes. The use of different and multiple counters in industrial applications will be discussed in this chapter.

THE BASIC PC COUNTER FUNCTION

Figure 13-1 shows the configuration of the basic PC counter function. The counter shown has a double input. A single input counter is impractical, because turning a single input off between pulses would reset the function to 0 after each pulse. A single input function would never get beyond a count of 1. Separate count and reset inputs are required as shown in the figure.

The PC counter functional configuration is almost identical for up counters and down counters, except, of course, one counts up and the other counts down. Figure 13–1 illustrates two typical formats for PC counters. The two inputs shown control the counter operation. If IN 0002 is open, the counter is set to 0. When IN 0002 is closed, the counter is enabled. Any time during operation that IN 0002 is reopened, the counter resets to 0.

When enabled, the counter will count once each time IN 0001 goes from open to closed. It does not count when IN 0001 goes from closed to open. Suppose we program a PC up counter to a preset count of 21. Starting at 0, the counter increments one number each time the input pulses on. When a PC count of 21 is reached, output CR 0017 will go on. As IN 0001 continues to be pulsed beyond 21, there will be no change in the output. It stays on and the counter continues to increment.

The down counter operates similarly to the up counter. In our example, we would start counting at 21. As input pulses are received at IN 0001, the counter increments downward (21, 20, 19, etc.). When the count reaches 0, the counter output will energize. Additional pulsed inputs to IN 0001 will have no further effect on the output status.

Unlike the PC timer operation, the counter operation is normally not retentive. Opening the enable input, IN 0002, at any time will reset the counter to 0 for a PC counter. When IN 0002 is reclosed, the count has

Figure 13-1
The PC Counter
Function

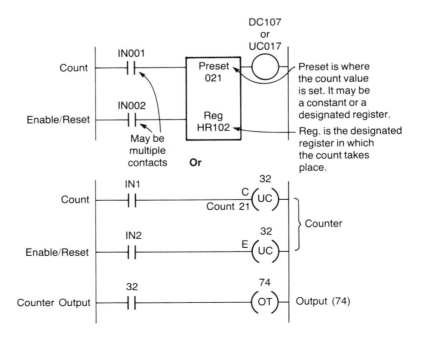

not been retained as a restarting point. Check the PC operational manual to see whether your PC counter function is retentive or not.

CHAPTER EXAMPLES

We will show three examples of the use of the PC counter. The first is the basic use for counting events. The second and third examples use more than one counter or control a process flow. The examples are:

1. Straight counting in a process. The counter output goes on after the set count is received by repetitive pulses to the counter input.
2. Two counters used with a common register to give the sum of two counts.
3. Two counters used with a common register to give the difference between two counts.

Example A in figure 13-2 illustrates the fundamental use of a PC counter. After a certain number of counts occur, the output goes on. The

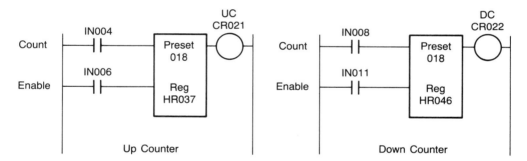

Figure 13-2
Example A. Basic PC Counter Operation

output can be used to energize an indicator. The output status could also be utilized in the ladder diagram logic in the form of a contact. The counter function is shown for either an up counter or a down counter. They both perform the same function in this illustration.

Either counter will function if its enable line is energized. After the count input receives 18 pulses, the CR output will energize.

Example B in figure 13-3 illustrates the use of a combination of two counters. Suppose we wished for an output indicator to go on when six of part C and eight of part D are on a conveyor. This circuit would monitor the proper counts. IN 0002 and IN 0003 are proximity devices that pulse on when a part goes by them. Note that the circuit would not indicate

Figure 13-3
Example B.
Dual Counter
Application

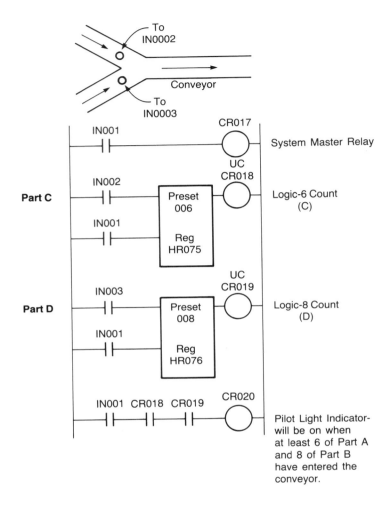

more than six or eight parts; it would only indicate when there are enough parts.

To repeat the process, turn IN 0001 off to reset the system. Then reclose IN 0001.

Example C, shown in figure 13-4, concerns keeping track of the net number of parts on a conveyor. The number of parts going on the conveyor is counted by one proximity device's count. The number leaving the conveyor is counted by a second proximity device's count. Each proximity device feeds information into its own counter function. The total net count is kept in a holding register common to both counters.

An accurate initial count is needed. When starting the operation the number of parts on the conveyor must be determined. This count number is programmed into the common register, HR 0101. It is normally

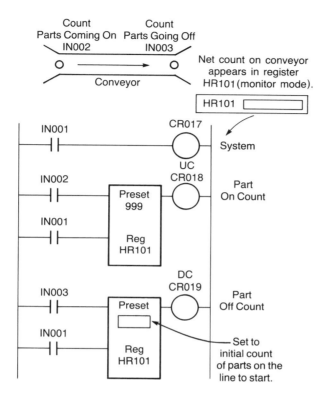

Figure 13-4
Example C.
Counters Used
for a Net Count

necessary to put this count number in the down counter as the preset number. Then, any parts going onto the conveyor pulse the up counter. The counter's register (which is common to both counters) will have its value increased by one for each entering part. Similarly, the parts leaving decrease the common register's count through the down counter. The number value in register HR 0101 represents the number of parts on the conveyor. We are assuming that no parts fall off the conveyor and none are added along the way.

The up counter preset value is irrelevant in this application. It does not matter whether the counter outputs are on or off. The output on-off logic is not used. We have arbitrarily set the up counter's preset values to the maximum.

PROGRAMS WITH BOTH A COUNTER AND A TIMER

There are many PC applications in which both the counter function and the timer function are used. This section illustrates three examples. The first example, D, is for a timed process that occurs after a certain process count is reached. After a count of 15 from a sensor, a paint spray is to

Figure 13-5
Example D.
Count and Time
Program

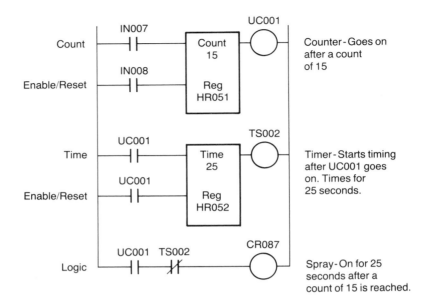

run for 25 seconds. Figure 13-5 shows a program that does the count and time operation.

Example E, shown in figure 13-6, is for a delayed counting period. In this process we do not wish to start counting until one hour after the process starts. A timer output contact in the timer run line closes after the time period. The closure then enables the counter to start counting input pulses. After a count of 150, the output comes on.

How many parts per minute are going past a certain process point? Example F addresses this problem. Figure 13-7 is a ladder diagram

Figure 13-6
Example E.
Delay of the
Start of the
Counting Process

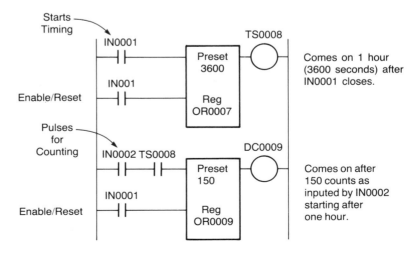

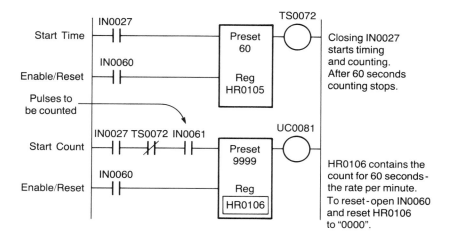

Figure 13-7
Example F.
Rate-Per-Time-
Period Program

scheme for obtaining the product part flow rate. The time and counter are enabled at the same time. The counter is pulsed for each part going past its sensor, which is connected to IN 0027. The counting begins and the timer starts timing through its 60-second time interval at the same time.

At the end of 60 seconds, the timer's count ladder line is opened by a normally closed timer-output-related contact. Pulses continue but do not affect the PC counter. The number of parts for the past minute are now recorded in counter register HR 0106. The part count for the minute can be recorded manually or by a computer technique and will remain in HR 0106 until IN 0060 is opened and the counter and timer are reset. After IN 0060 is reclosed, the process starts over.

EXERCISES

Design, construct, and test PC circuits for the following processes:

1. An indicating light is to go on when a count reaches 23. The light is then to go off when a count of 31 is reached.

2. A stacking/banding system (S) requires a spacer to be inserted (I) in a stack of panels after 14 sheets are stacked. After 14 more (28 total), the stack is to be banded (B). Add sensors and assumed output devices as required.

3. Refer to exercise 2 and add the following additional steps to the process. After banding is completed, there is a two-second delay for the bander to pull back. Then, an identification spray color dot (P) is to be applied to the stack. Spray time is four seconds.

4. There are two feeder conveyors (F1 and F2) feeding a part onto one main conveyor (M). There is a proximity device at the end of each feeder conveyor. The proximity device outputs are fed as pulses to counters. Each counter the shows the count of parts being put on to the main conveyor. In addition, another proximity device at the end of the conveyor pulses in response to parts leaving and then sends the pulses to another counter.

Develop a program to have a single register showing the number counter of parts on the conveyor. Assume that the register is initially set to the same count as the count of parts on the conveyor.

5. Program an automatic control for the system shown in figure 13–8.

Figure 13–8
Diagram for
Exercise 5

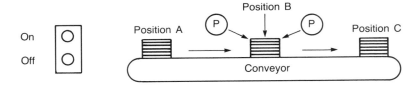

When the on button is pushed, a stacker (S) starts stacking plywood sheets at point A. Stack height is controlled by a PC counter function, not a height sensor. When 12 parts are stacked, the conveyor (CV) goes on and moves the stack to point B. A sensor is used to stop the conveyor at B. At B, paint (P) is applied for 12.5 seconds. After painting is complete, the conveyor is restarted manually. The conveyor then moves parts to point C. At point C the stack stops automatically and the stack is removed manually. The stop button stops the process any time it is depressed. Assume that only one stack is on the conveyor at a time. Add limit switches, etc. as required.

Auxiliary Commands and Functions

14

At the end of this chapter, you will be able to

☐ Explain how the MONITOR mode may be called up and used for ladder diagram analysis.

☐ Explain how the FORCE mode is used for PC program testing and analysis.

☐ List the safety precautions required when using the FORCE mode.

☐ Explain how the PRINT mode is used to print out ladder diagrams.

☐ List and explain the other major types of PC PRINT capabilities.

INTRODUCTION

Three important PC functions deserve a separate chapter to cover their usefulness: the MONITOR mode, the FORCE/OVERRIDE function, and the various PRINT capabilities.

After a circuit is programmed into a PC, its operation may be watched on a screen in the MONITOR mode. The current flow from left to right as contacts open and close is indicated by a brightening of the screen pattern. Functions such as coils and timers also light up when they become energized. Other types and models of PCs show the current flow by flashing lines and functions. Still others use a dotted pattern system. This chapter will discuss the use of the MONITOR mode.

The second function to be covered in this chapter is the FORCE function. In some cases, this function may be looked upon as an override control. To use the FORCE function, first call up the contact, coil, or function to be controlled. Next, the cursor is moved to the function to be controlled. Then, the FORCE function key is depressed. Then, using the keyboard keys, the function under FORCE control may be turned on and off. The keyboard then overrides the status of the input from the outside system.

The third function to be discussed is the use of printouts to record information regarding a circuit and the status of the circuit parts. The most common printout is that of the ladder diagram. There are also other printouts available on many other PC models. These are for registers, timed status information, and other PC equipment status.

MONITOR MODE—LADDER DIAGRAMS

The MONITOR mode for ladder diagram operation is indicated on the screen in various ways. It may be indicated by a brightening of the pattern where voltage is passed through. In other cases it is indicated by the pattern changing to a dotted line or to a flashing effect. A large monitor that shows complete ladder lines normally uses the brightness enhancement effect. Smaller monitors showing a portion of a line use the other indicating systems. Figure 14-1 illustrates brightness enhancement for a standard, three-wire, motor control, single-line ladder diagram. The figure shows the screen as the two inputs (stop and start) are energized and de-energized. The pattern changes allow us to watch the circuit operation.

The MONITOR function is especially useful for analysis of a large number of ladder lines. The MONITOR mode assists the operator in troubleshooting a large system that is malfunctioning. In some PC models the screen is in MONITOR mode whenever it is in the EDIT mode; in other PCs, the MONITOR mode must be called up separately.

Figure 14-1
MONITOR
Mode Example

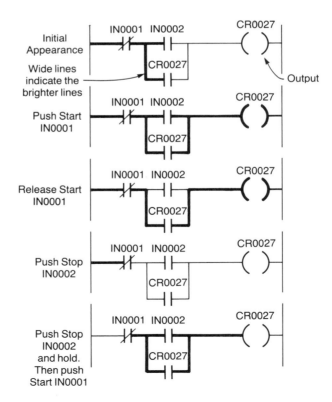

Other MONITOR Mode Functions

There are other system characteristics that may be monitored in addition to the ladder diagram. These include register status (value), as well as individual coil and contact status. Other monitorable system parameters include a listing of the forced functions, which will be discussed next. Some advanced PC systems can also list the actual malfunctioning output devices for fast analysis.

Figure 14–2 shows how the status of four holding registers would be shown. This figure shows the register values in binary; many PCs give you a choice of which numbering system you want used for the printout: decimal, hex, octal, or ASCII, as well as binary.

In most PC systems, you may call up individual coils, contacts, or both on the screen. For example, if you are looking at or in the vicinity of line 32, you may wish to see what is happening to an input contact on line 6, which is off the screen. Contacts and coils from line 6 can be inserted in a blank space by themselves and observed for on-off status. Figure 14–3

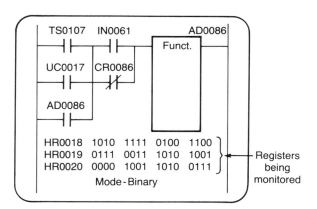

Figure 14–2
Individual
Register Status
Display

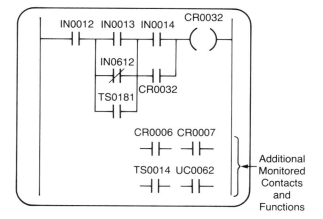

Figure 14–3
Contact Status
Monitoring

shows how these individual contacts might appear on a screen along with other PC information, such as the ladder diagram.

THE FORCE MODE

Many PCs have the capability to carry out a FORCE function. The function is essentially an override control that enables the operator or programmer to operate the circuit from the program keyboard.

The FORCE procedure is normally carried out in the MONITOR Mode. First, place the cursor over the contact, coil, or function you wish to force. Then carry out the specific keyboard procedures for forcing. Turning the FORCE on changes the status of the contact or coil under the cursor. If it is a normally open contact, it will close (turn on). If it is a normally closed contact, it will open (turn off). If you force a coil or function, it will go on when forced, regardless of external commands in effect.

In most cases, forcing any contact of a relay, CR, will turn the relay coil on, as well as forcing all other contacts of that CR number. An example of a display showing a FORCE procedure is shown in figure 14-4. To remove the FORCE function, turn it off and then press the Clear key.

The individual coils or contacts that are forced on may be left in the forced state permanently by entering them—usually by pressing return.

Figure 14-4
Force Procedure

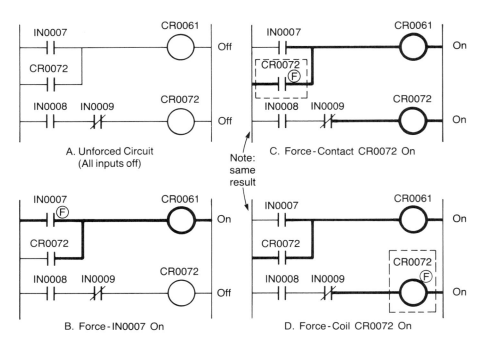

This permanent-entry procedure must be done carefully so as not to introduce a permanent unwanted change in an operational sequence.

There are certain limitations of the FORCE mode. Not all functions react like coils and contacts when forced. For coils and contacts, forcing a contact causes its coil and all of its other contacts to be forced at the same time. Many other functions work in the same manner, but not all. It is necessary to review the operating procedures of a particular PC to see how each function responds to the FORCE command.

For example, in some PCs, forcing the MASTER CONTROL RELAY (MCR) function does not have the same effect as officially turning the MCR function on through its normal operating ladder program. Forcing the MCR coil does not affect its function but does turn on its associated contacts.

The same special consideration can apply to the SKIP and DR/SEQUENCER functions. See your operating manual for individual function FORCE operational characteristics.

If an industrial process is in operation, it obviously would be undesirable to insert into it periodic false signals by hooking up a keyboard to the CPU controlling the process and forcing in the false input signals. Not only would this be dangerous to equipment but, more important, someone could be injured. Do not use the FORCE function on an operating system unless all personnel in the area have been notified. It is best not to use it at all on an operating system; limit its use to simulations, if at all possible.

PRINTING LADDER DIAGRAMS

Ladder diagrams on a screen cover from one to four or five rungs, depending on the PC model. If the entire operational circuit has twenty or more rungs, for example, you may wish to see the entire circuit at once. If the PC you are using has a PRINT mode system, the whole ladder diagram can be printed out continuously on a conventional computer printer.

There are, of course, other reasons you might want a ladder printout. You might need a permanent written record, for instance. Also, in education and training, a printout is a written record of laboratory achievement (under proper controls).

One helpful extra feature of many ladder printouts is that each rung may be printed with a cross-reference system. These cross references are similar to the conventional ones used in standard ladder diagrams. Each ladder line with a coil or function is assigned a consecutive number. Then, on each line, a listing is printed of the other lines in which contacts from that line's coil or function occur.

Figure 14–5 is an illustration of a PC ladder printout. The cross-reference system is included as numbers referencing other sections.

Figure 14-5
Ladder Diagram
Printout

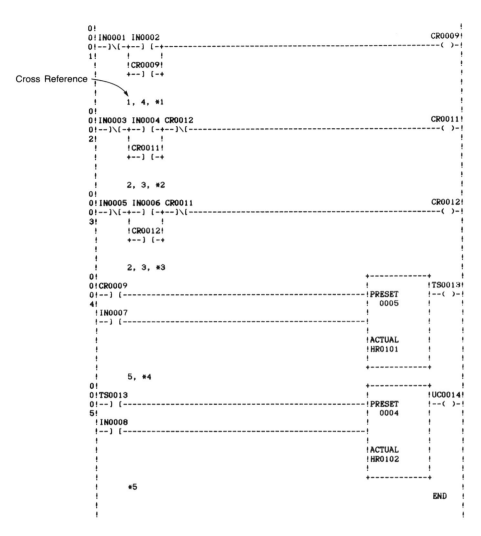

Other PC PRINT Capabilities

Other typical PRINT mode capabilities include register status, FORCE mode status, timing diagrams, input status, output status, and a listing of malfunctioning output devices. This section will discuss only register status, FORCE mode status, and timing diagrams.

A register status printout is shown in figure 14-6. A user-friendly screen program lets you choose to have the status of the register, consecutive registers, or a number of nonconsecutive registers printed. In most cases, you may choose what numbering system you want the printout to display: binary, hex, octal, decimal, ASCII, or others, depending on your PC model.

IR0706	1010	1111	1100	0000
IR0707	0101	0010	0110	1101
IR0708	1001	0101	0011	1111
IR0709	1100	1000	1010	1101

(In Binary)

HR0062	0087
HR0063	0642
HR0064	7410
HR0065	0007

(In Decimal)

Figure 14-6
Register Status
Printout

The printout may be in hard copy form from the printer, or it may be printed on the monitor screen.

The status of any forced functions may also be printed out. If you have forgotten what forced contacts or functions remain in the ladder diagram, a FORCE listing printout will display them. If there are no forced contacts or coils, none will print out; otherwise, those in effect in the circuit will be printed out. Like the register status, the FORCE listing may be put on a printer or printed on the screen.

Timing diagrams are available on many PC printouts. In most, you first choose the time interval you wish to use, from tenths of a second to minutes. Next, you choose the item or items to be observed, such as registers or coils and contacts. The number of items viewed is limited by the PC's program and printer column width. Figure 14-7 illustrates the printouts for two registers being timed. Figure 14-8 shows five contacts (each with the same number as its coil) being timed. Both figures shown use a fixed, selected interval.

Time	HR0307	OR0072	IR1072
0	0682	0167	6421
5	0682	0268	6421
10	0683	0167	6421
15	0683	0167	6421
20	0684	0268	6421
25	0685	0411	6421

Set for a fixed 5-second interval (in decimal)

Figure 14-7
Register Timing
Printout

Time	CR0121	CR0071	CR0006
0	101		
2	101		
4	111		
6	001		
8	101		
10	000		
12	001		
14	011		

Key

1 0 1
Means:
CR0121 - 1 - On
CR0071 - 0 - Off
CR0006 - 1 - On

Set for a fixed 2-second interval

Figure 14-8
Contact Timing
Printout

An alternative to fixed time intervals is available. Exception time saves paper and the time that would be spent poring through a lot of data. Exception timing prints out only when one of the items being monitored changes status. Figure 14–9 shows how exception time works for the same registers and contacts shown in figures 14–7 and 14–8. The time of the status change is shown on the left of the printout.

Timing in intervals and by exception can be shown on the screen or printed on a printer.

Figure 14–9
Exception
Timing Printout

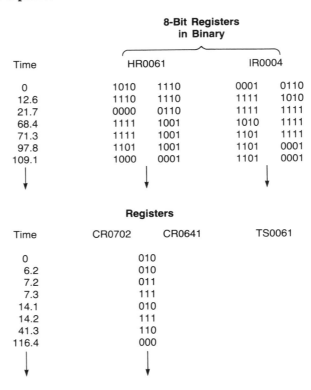

EXERCISES

Obtain the operational manuals for one or more PCs and review the operational procedures for MONITOR, FORCE, and PRINT. Answer the following questions:

1. Explain how the MONITOR function is made operational, how it works, and what data and functions may be observed.
2. Repeat exercise 1 for the FORCE function.
3. For the FORCE function, list how the force procedure affects each of the operational functions of the PC, starting with contacts and coils.
4. Repeat exercise 1 for the PRINT function.

SECTION FOUR

INTERMEDIATE FUNCTIONS

Arithmetic Functions

15

INTRODUCTION

Medium and large PCs have arithmetic function capabilities. This chapter will cover six of these arithmetic functions and their applications.

The functions to be explained are ADDITION, SUBTRACTION, MULTIPLICATION, DIVISION, SQUARE, and SQUARE ROOT. There are many processes that need these arithmetical operations on a fast, continuous basis. The PC can do many arithmetic operations per second for fast updating when needed. The usual interval between PC arithmetical function updates is one or two scan times.

Chapter 15 will also show where DOUBLE PRECISION, the use of two adjacent registers, is used. Multiplying 2 four-digit numbers together results in an eight-digit number, which will not fit into a four-digit slot. Two slots, or registers, are used automatically by the PC for the resulting number. This system and other operations requiring double precision will be illustrated, along with the PC handling of negative numbers for each of the functions.

ADDITION

Figure 15–1 shows the PC ADDITION function. It adds only when the enable line changes from off to on. The addition will not take place continuously just because the enable is on. When enabled, the numerical value

Figure 15–1
The PC ADDI-
TION Function

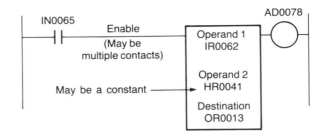

in the operand-2 register is added to the numerical value in the operand-1 register. The resulting value then appears in the specified destination register. In many PCs, operand 2 can be a number or the value in a designated register.

When does the output coil come on? In some PCs its status is irrelevant. In other PC systems, the coil's energization indicates register overflow or negative value. For overflow, the coil only comes on when the resulting number exceeds the register counting capability; otherwise, it remains off. For example, assume the decimal register limit is 9999. If you add 643 plus 568, the sum, which equals 1211, is within the register limit, and the coil will not come on. The sum, 1211, will appear in the destination register. On the other hand, the sum of 8973 plus 8632 is 17,605, which exceeds 9999, so the coil will come on. In this case, the destination register will read 7605, the excess over 10,000.

Note that we are using decimal numbers throughout this chapter. The arithmetic functions also work for binary numbers and other numbering systems, as well. You may choose any numbering system, depending on the requirements of your particular PC model. The only precaution is to use the same numbering system throughout the arithmetic function.

To find out which register types, HR, OR, etc., can be used as operands, consult the manufacturer's manual for your PC.

Figure 15–2 illustrates the ADDITION function for two examples. One example is for a result less than 9999, and the other is for a result greater than 9999.

A sample industrial problem for the ADD function is shown in figure 15–3. Two conveyors feed a main conveyor. For some reason we cannot get to the main conveyor to make a count. The main conveyor count is determined from the count of parts entering from the other two conveyors. The count on each feeder-conveyor is determined by a counter (not shown). The counters on each feeder-conveyor are input-pulsed by a proximity detector once for each part leaving the conveyors. The count of total parts entering the main conveyor is then determined by adding the two feeder conveyor counts using the ADD function. For illustration, we monitor the total count every 30 seconds. The input of the ADD function is pulsed

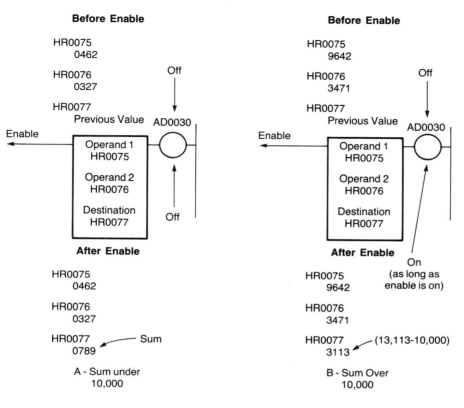

Figure 15-2
Two
ADDITION
Examples

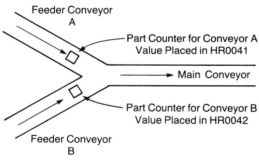

Figure 15-3
Using the ADDI-
TION Function
For a Conveyor
Part Count

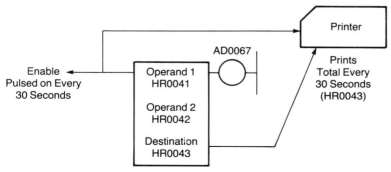

on and immediately off by pulsing the ADD function enable. The count could be printed out as shown in figure 15–3 every 30 seconds.

Note that this addition method is an alternate solution to a similar problem in chapter 13. Chapter 13 used counters and a common register.

SUBTRACTION

The PC format for SUBTRACT is the same as for ADD, and the function operation is similar. For SUBTRACTION, operand 2 is subtracted from operand 1. The result is found in the destination register. Figure 15–4 shows the SUBTRACT function.

Figure 15–4
The PC
SUBTRACTION
Function

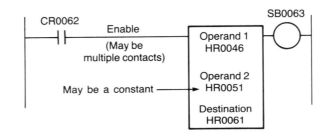

Figure 15–5
Two SUBTRAC-
TION Examples

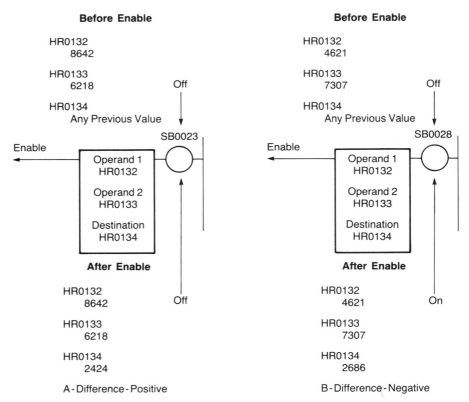

As with addition, the SUBTRACT function operates only when the enable line goes on. In some PCs, the direct answer will appear. In other PCs, the coil status is significant for a complete answer description. The coil on-off operation for SUBTRACTION differs from that of ADDITION. When the result is positive, the coil is off and the result is found in the destination register. When the result is negative, the coil is on and the resulting negative number value is found in the destination register.

Figure 15–5 shows the function operation for both a positive and a negative answer. The coil is off for a positive result and on for a negative result.

A possible industrial problem for SUBTRACTION is shown in figure 15–6. It is similar to the ADDITION example. In this example, the out-

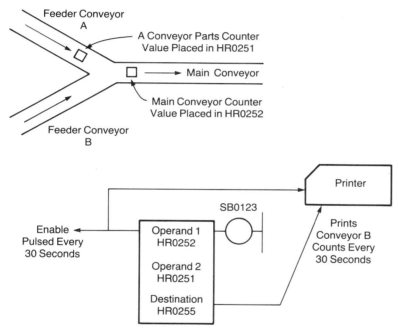

Figure 15–6
Using the SUB-
TRACTION
Function for a
Conveyor Count

put count and only one input count are available. One of the conveyor inputs is not accessible for some reason. To obtain the input A count value, subtract the input B count value from the output count. The result is the A conveyor count. The count is again determined by a 30-second counting interval. The count is taken for 30 seconds, used, and then reset to zero. Again, initializing is needed periodically for accurate operation. Initialization figures the initial number of parts on conveyor B into the counting results.

THE REPETITIVE CLOCK

The ADD and SUBTRACT functions discussed so far in this chapter do not usually operate continuously. They only perform the addition or subtraction operation once when the function is enabled. If enable remains on, nothing else happens, even though the operands change. When enable is off, nothing happens, either. Therefore, a repetitive on-off enable is needed for continuous operation.

Figure 15–7 shows a repetitive clock arrangement. A coil turns itself off and on at a very fast rate, about twice the scan-time rate. If this is used as an enable, the operation of an arithmetic function is essentially continuous—if you consider every millisecond or so to be continuous.

Figure 15–7
Repetitive Clock
Circuit

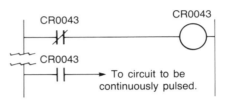

The sequence for the repetitive clock is: On the first scan, the relay coil is turned on through its own contact. When the coil goes on, it opens its own enabling contact, which is normally closed. At the end of the first scan, the CR 0043 NC contact is updated and opened. On the next scan, the coil is turned off. Update at the end of the second scan recloses the self-enabling contact. The process then repeats itself continuously.

When fixed, longer intervals are needed, timers are used instead of the repetitive clock. These interval situations were covered in chapter 12.

Using the ADD and SUBTRACT Functions to Set a Range

In this example, which will set a range using the ADD and SUBTRACT function, there is an inspection system with a periodically changing base dimension and periodically changing tolerances. The PC can easily and quickly reset the dimensions and tolerances. The preset dimensional values are transmitted to two specified PC outputs. These output values are used to set the positions of an automatic gauging system.

In the example, the base dimension, or set point, is set at 6.250 inches. The allowable tolerances chosen for this illustration are +0.025 and −0.035. Figure 15–8 shows the range in graphic form.

Figure 15–9 illustrates how the settings can be accomplished by programming a PC. For three eternally fixed values, a PC is not needed. Our system is valuable when the dimensions are varying quickly as production varies. The set point is entered into HR 0001. HR 0002 receives the upper tolerance value, and HR 0003 receives the lower tolerance value. When the circuit in figure 15–10 is enabled, the upper and lower limits are calculated. They appear in HR 0004 and HR 0005, respectively.

To revise dimensions or tolerances, the mathematical values in HR 0001, 0002, and 0003 are changed. When PC functions are re-enabled, the

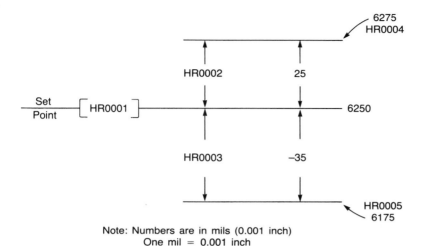

Figure 15-8
Graphic
Representation
of Dimensions

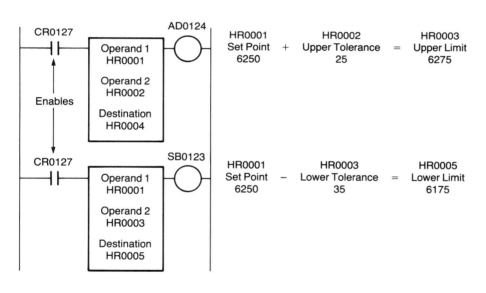

Figure 15-9
PC Operation
with Set Points
and Tolerances

new values appear immediately (on the next scan) at the PC output. Comparing the actual dimensions with the upper and lower limits we have set will be discussed in chapter 16, which covers COMPARISON functions. Moving different dimensional numbers into the set point and tolerance registers will be discussed in chapter 18, which covers the MOVE functions.

MULTIPLICATION

The MULTIPLICATION format is similar to the ADD and SUBTRACT formats previously discussed. Figure 15-10 shows the MULTIPLICA-

Figure 15-10
The MULTIPLI-
CATION
Function

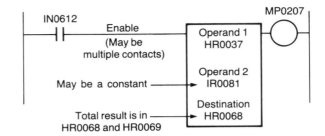

TION function. Operand 1 is assigned a register number. Operand 2 can be another register, or it may be a constant. The result of the multiplication appears in the destination when the function is enabled. The destination is two registers wide by necessity. Multiplying 0034 by 0086 would require only one register, four numbers wide, for the answer (2064). However, multiplying 6453 by 8933 (57,644,649) would require an eight-bit-wide slot or two registers to accommodate the answer.

The multiplication takes place only when enable comes on. Normally, the coil comes on when the multiplication is completed. The coil on-off state has no numerical significance as it did in ADD and SUBTRACT.

A simple process application for counting cartons is shown in figure 15-11. The count from a carton counter enters the PC and is put into IR

Figure 15-11
MULTIPLICA-
TION Example

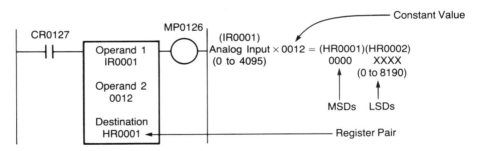

0001, and then into operand 1. Each carton contains 12 bottles; therefore, 12 is entered as a constant in operand 2. To keep a constant count of bottle output, the PC would constantly multiply the carton count by 12. The destination register, HR 0001, will show the number of bottles output each time the function is enabled. A number of these MULTIPLY functions could be combined to give a total plant unit output. The constants in each function would be bottles per carton for that particular count. A PC addition program of all individual counts would then give the total plant bottle count.

SQUARING

There is normally no SQUARING function in a PC format. Squaring is simply a matter of putting the number to be squared into both operand

1 and operand 2 of a MULTIPLICATION function. The square of the original number then appears in the destination register. The SQUARING function is shown in figure 15–12.

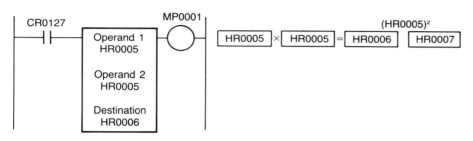

Figure 15–12
Squaring by the
MULTIPLICA-
TION Function

DIVISION

The DIVISION function, which is shown in figure 15–13, is similar to the MULTIPLICATION function. Operand 1, the dividend, is divided by operand 2, the divisor. The numerical result of the division appears in the destination register when the function is enabled. Again, the division takes place only at the time the enable is energized. To facilitate division, operand 1 is two registers wide, and operand 2 is only one register wide. Operand 2 may normally be a register or a constant numerical value.

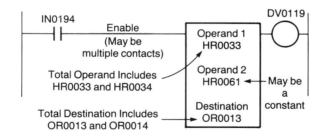

Figure 15–13
The DIVISION
Function

In PCs, the destination is almost always two registers wide. The first destination register is the numerical result of the division. The second register value is the remainder in numerical form. The remainder is not a decimal value. A numerical example of a division is illustrated in figure 15–14. The determination of number value in the second destination register is explained in the figure.

Figure 15–15 is an example of scaling by division. An analog measurement numerical value, in inches, is fed into a PC input register. The measurement value is transferred (not shown) within the PC to input register IN 0078. To get the value in feet to output from the PC into an indicator, divide by 12 with a DIVIDE function. The result, now in the required dimension of feet, appears in register OR 0124.

Figure 15-14
DIVISION Destination Register Content Determination

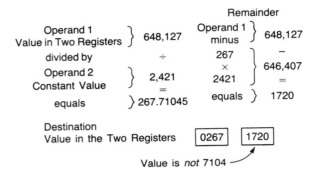

Figure 15-15
Example of the Process Use of the DIVISION Function

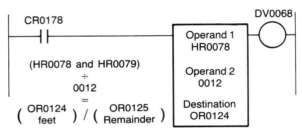

If the required output accuracy is less than whole feet, convert the remainder into decimal form. This would be accomplished by an added program step (not shown): divide the numerical remainder in OR 0125 by 12 and place the result in another output register, for example, OR 0126. You would then have four-decimal-place accuracy when the OR 0126 output register value is recognized as ten-thousandths of an inch.

SQUARE ROOT

Figure 15-16 shows the SQUARE ROOT function found on some PCs. The number whose square root we want to determine is placed in the source. The source input number is contained in two registers so that it may be up to 99,999,999 in value. When enabled, the function calculates the square root and places it in the destination. The destination is one register wide, up to 9999 in value. There is usually no remainder register.

Figure 15-16
The SQUARE ROOT Function

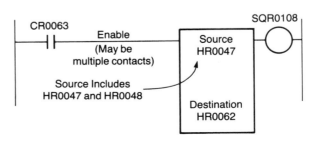

DOUBLE PRECISION

Suppose a process needs seven- or eight-place figure accuracy, not just the normally calculated four figures. Some advanced PCs have the capability to double the number of output decimal digits, for example, from four digits to eight. For more precise processes, this increased accuracy may be required. The PC's system to increase accuracy is called DOUBLE PRECISION. Figure 15–17 illustrates one manufacturer's system for carrying out DOUBLE PRECISION for the ADD function. Consult your user's manual for how your particular PC does this, or, indeed, whether it can do it at all.

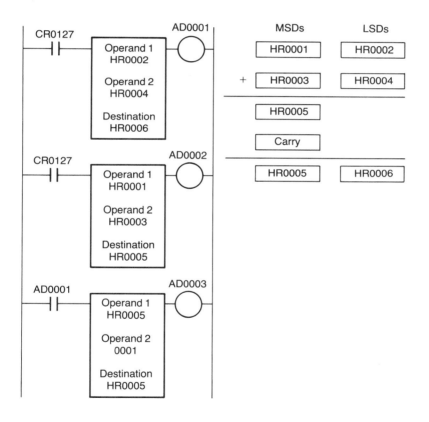

Figure 15–17 Typical DOUBLE PRECISION Function Format

EXERCISES

All exercises assume that all numbers are in decimal form.

1. Construct a basic PC ADD function. Use IN 0001 as the enable circuit. Use HR 0001 and HR 0002 as the registers holding the numbers to be added and HR 0003 as the sum. Insert relatively small num-

bers in HR 0001 and HR 0002, enable the function and verify that the resulting sum in HR 0003 is correct. Use the appropriate numbering system to accomplish the addition.

Next, insert large numbers whose sum exceeds the capability of HR 0003 and observe the result. Does the result correspond to figure 15–2?

2. There are three conveyors feeding a main conveyor. The count from each feeder conveyor is fed into an input register in the PC. Construct a PC program to obtain the total count of parts on the main conveyor. HINT: add the first two conveyor counts and then take that sum and add the count of the third to it in a second function.

As an additional exercise, use a timer to update the total every 15 seconds.

3. Construct a SUBTRACT function in a manner similar to exercise 1. Insert numbers in the operands that result in a positive destination number. Next, use numbers that give a negative answer. Does the negative answer produce the results as shown in figure 15–5?

4. Two conveyors, A and B, feed a main conveyor, C. A third conveyor, R, removes rejects a short distance down the main conveyor. The counts for conveyors A, B, and R are each input into holding registers in the PC. Construct a PC program to obtain the total output, C, part count. HINT: use two steps—one addition and one subtraction.
As an additional exercise, use a timer to update the total at a time interval of your choice.

5. Set up a process-range PC program following the system shown in figures 15–8 and 15–9. The nominal value or set point is 15.35 inches. The tolerances are +0.27 and −0.27 inches. Show that the resulting PC-calculated limits are correct. To further test the program's validity, change the set point and tolerances to different values and check the results.

6. Repeat exercise 5 with different tolerances. Both tolerances are negative, −0.05 to −0.20.

7. A main conveyor has two conveyors feeding it. One feeder puts six-packs on the main conveyor and the other feeds eight-packs. Both feeder conveyors have counters that count the number of packs leaving them. Design a program to give a total can count on the main conveyor. HINT: two multiplications and one addition are needed.

8. A conveyor has 6-, 8-, and 12-packs of canned soda entering it. Each size of entering pack has an individual pack quantity counter feeding a PC register. To know how many total cans are entering the conveyor, set up a program for multiplying and then adding to give a total can count. HINT: three multiplications and two additions are needed.

9. We have an output that gives us a dimension in inches. We wish to have the dimension displayed in feet and yards. Develop a PC program to output all three dimensions. HINT: use two divisions for two outputs and one direct-access output.

10. Set up a PC program to obtain an output, P, in register OR 0055. The output is to give a value based on two inputs, M and N. P equals the square of M plus the square root of N.

11. Develop programs for other math equations of your choice. Example: N = (J + K − L)/M.

Number Comparison Functions

16

At the end of this chapter, you will be able to

☐ List and define the six COMPARE functions.
☐ Show numerically the capabilities and application of each of the six comparisons.
☐ Apply each of the six numerical COMPARE functions to solve an applicable process problem.
☐ Apply combinations of COMPARE functions to do multiple comparison analysis.

INTRODUCTION

Medium and large PCs have number comparison function capabilities. The number comparisons are performed internally in a manner similar to microcomputers and microprocessors. With the PC, there is no internal programming necessary for the operator. The PC programming is set up for direct keyboard/screen arithmetical logic. Chapter 16 will illustrate how to perform number comparisons of all types of PC programming.

What kind of number comparisons can be made by a PC? We may wish to compare two numbers. We might compare a varying count to a fixed value. We might wish to compare two varying input values every five seconds.

In an even more complicated situation, we might wish to determine whether a periodically varying number is between two limits. These limits of comparison might be fixed, or one or both limits could be variable.

In this chapter we shall see how PC COMPARE functions work, and how they may be applied to numerical processing problems.

THE SIX COMPARISON FUNCTIONS

Many PCs have only two COMPARE functions: equal, and greater than or equal to. To perform any one of the other four functions (not equal, less than, greater than, and less than or equal to) combinations of the basic two are used. Some PCs have all six individual functions, which makes programming easier. Of course, some less-expensive PCs do not have COMPARE functions at all.

This chapter will use the PC comparison system with two basic functions for illustrations. The other four PC COMPARE functions will use the inverse or combinations of the two basic functions.

Figure 16–1 shows a table of COMPARISON functions. Functions 1 and 3 are the two basic functions that we have discussed. The other four are derived functions.

Let's take an example of each COMPARE function. Assume that A, the standard for comparison, is placed in operand 2. A is set at 182. Then B, the numbers to be compared to A, will be placed in operand 1. We are therefore comparing the value of B to the value of A, 182.

1. Equal (EQ) is true only if B is exactly 182 also.
2. Not equal (NE) is true if B is 181 or less, or if B is 183 or more.
3. Greater than or equal to (GE) is true only when B is 182 or less.
4. Less than (LT) is true only when B is 183 or more.
5. Greater than (GT) is true only when B is 181 or less.
6. Less than or equal to (LE) is true only when B is 182 or more.

In actual operation, A might be a varying number, not a fixed value of 182. Later chapter examples illustrate how it may be changed periodically.

Figure 16–1
The Six
COMPARISON
Functions

Comparison	Function	Equation	Circuit (conducts when equation is true)
*1	Equal (EQ)	$A = B$	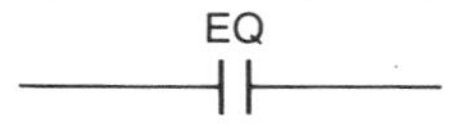 EQ ⊣⊢
2	Not equal	$A \neq B$	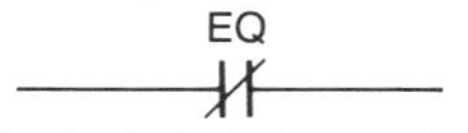 EQ ⊣/⊢
*3	Greater than or equal to (GE)	$A \geq B$	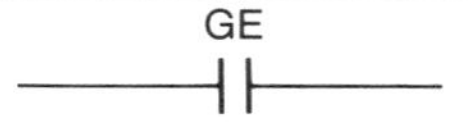 GE ⊣⊢
4	Less than	$A < B$	 GE ⊣/⊢
5	Greater than	$A > B$	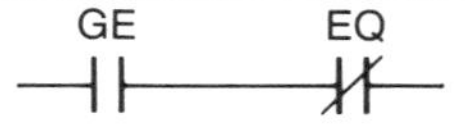 GE ⊣⊢ EQ ⊣/⊢
6	Less than or equal to	$A \leq B$	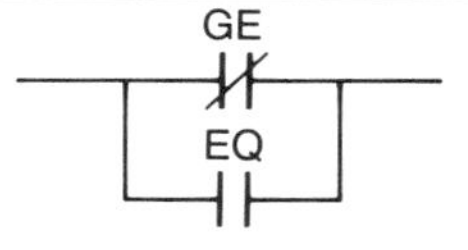 GE ⊣/⊢ / EQ ⊣⊢

* Basic Functions

THE GENERAL COMPARISON FUNCTION

Figure 16-2 shows two basic COMPARISON function layouts. The two numbers being compared are operand 1 and operand 2. One operand can be a constant and the other operand a register. Both operands may also be registers that contain numerical values. The identification number of the register would be specified in the functional block.

When the function is enabled by the input contact, the comparison is made. If the comparison is true, the output goes on. If the comparison is not true, the output goes off, or stays off. The comparison in some PCs is made continuously as long as the enable is on. It makes the comparison on each scan. In some other PCs the comparison is made only at the time the enable goes on. To make another numerical comparison, the input must go off and then back on.

The patterns of the two basic COMPARE functions are normally similar. Figure 16-2 shows two formats of layouts that may be used for any COMPARE function. The only difference between the two basic functions would be the coil designation (and the mathematical manipulation by the PC CPU).

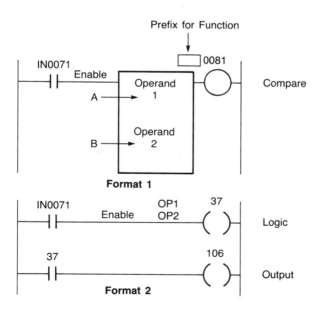

Figure 16-2
Typical PC
COMPARE
Function

CHAPTER EXAMPLES

This section illustrates the use of each of the six COMPARE functions in PC process control situations. Additionally, it shows one process application with two COMPARE functions used in combination.

Example A consists of two illustrations of the use of the equal-to COM-PARE function. We are banding dowels into bundles of 40. A counter function (not shown) keeps track of the count of the number of dowels in the bundle as they are added. The dowel counter's count number is kept in HR 0005. The running count is compared to 40, as shown in figure 16–3. When the count reaches 40, the comparison is true, and the output, CR 0019, goes on. Output CR 0019 is connected to a bander which operates when 40 is reached. The bundle removal and system reset systems are not shown. The other system would involve other functions.

The count in IR 0071 must be in the correct numbering system. We are comparing to a decimal 40, so IR 0071 must be in decimal for proper operation. Some PCs convert to the proper numbering system automatically and some do not. Appropriate number conversions may be needed as shown in chapter 9.

Another equal-to application is also shown in figure 16–3. In this case, an output must go on when two numbers are equal. The number's values do not matter, except that they are equal. The two numbers to be compared are fed from the outside process into HR 0123 and HR 0147 for the illustration. When enabled, output CR 0101 will come on any time the numbers in the two registers are exactly equal.

Example B is for the not-equal-to function. Figure 16–4 shows its programming. In the example, the output is to be on except when an input count is exactly 87. The input count is tracked in IR 0062. Operand 2 can be programmed as the number 0087. It could also be programmed as a

Figure 16–3
Example A.
Equal-To
Function

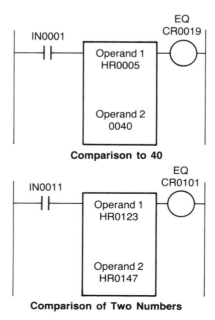

Comparison to 40

Comparison of Two Numbers

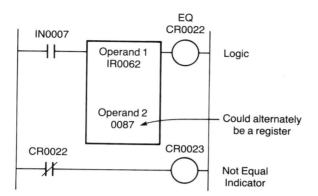

Figure 16-4
Example B.
Not-Equal-To
Function

register, HR 0183. The number 87 would then be inserted into HR 0183.

An automatic pill bottle filling operation has two possible bottle sizes. One bottle is to be filled to a count of 225 or more. The other is to have 475 or more. This example, C, uses a greater-than-or-equal-to COMPARE function. Figure 16–5 shows the PC function to control the pill counts. The pill count (counter not shown) is fed from an input to IR 0142 as the bottle is filled. The appropriate minimum number of pills for proper filling, 225 or 475, is inserted into HR 0128. A bottle is put under the pill dispenser (not shown). For a small bottle, the 225 limit is put into HR 0128; for a large bottle, 475 would be entered into HR 0128.

As the bottle starts filling, enable is continuously pulsed. The comparison is untrue and output CR 0030 is off. Once the pill count reaches 225 for the small bottle, CR 0030 goes on. Output CR 0030 is connected to a cap-and-remove operation (not shown). The bottle is capped, removed, and the process is reset and can be repeated. The same sequence would be carried out for the large bottle with HR 0128 set at 475.

Why not use an equal-to function for example C? EQ would probably work, but what if the process overshoots? Suppose the count somehow got to 226 for the small bottle. The fill would go on unabated. If the count

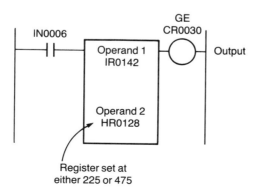

Figure 16-5
Example C.
Greater-Than-or-
Equal-To
Function

Figure 16–6
Example D.
Less-Than
Function

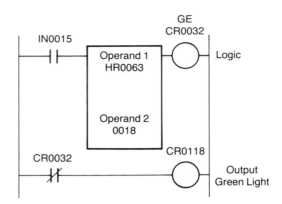

got to 226, or 476 for the large size, however, the fill process would not erroneously continue if you use the GE function.

Example D is illustrated in figure 16–6. Completed assemblies flow off of a production line. If an assembly is removed for rework somewhere along the line, the part to be reworked is automatically counted. If there are more than 18 assemblies removed in an hour, a light in the foreman's office is to turn off. If there are fewer than 18 rejects per hour, the light will remain on. The required hourly reset-to-0 system is not shown. (The only part of this program that is shown is the comparison portion.) The defect count is kept in HR 0063, operand 1. The allowable hourly number of defects, 18, is inserted as operand 2.

The greater-than function, Example E, is illustrated in figure 16–7. Two COMPARISON functions are required for this example. This production operation requires a count greater than 348 for the output to turn on. The number 348 is inserted as operand 2 in both functions. For a count of 347, the EQ function keeps the output off. Below a count of 348, the GE function keeps the output off. For 349 or more, the output is on as both contacts will be closed for the indicator ladder line.

The less-than-or-equal-to COMPARE function, Example F, is shown in figure 16–8. A production system produces a product that can be one of three colors: red, white, or blue. The production is limited to 348 blue units per day. The blue units are counted by using a color-sensitive detector. The detector count is fed to the PC into HR 0111. The maximum desired count, 348, is inserted into HR 0012.

The indicator is on for counts below 348. When the count reaches 348, GE stays on, and EQ comes on. The output remains on. When the count goes up one more to 349, EQ goes off, indicating that the production limit for blue has been reached. The output is now off and will remain off for higher counts.

The final example, G, shown in figure 16–9, is a multiple-compare program for lighting an indicator only when the count is between 15 and 22.

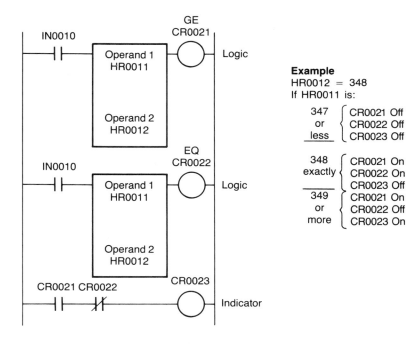

Figure 16-7
Example E.
Greater-Than
Function

Example
HR0012 = 348
If HR0011 is:

347
or
less
{ CR0021 Off
CR0022 Off
CR0023 Off

348
exactly
{ CR0021 On
CR0022 On
CR0023 Off

349
or
more
{ CR0021 On
CR0022 Off
CR0023 On

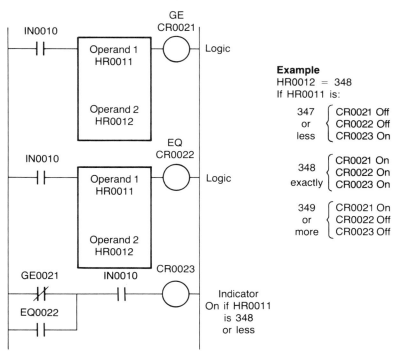

Figure 16-8
Example F.
Less-Than-or-
Equal-To
Function

Example
HR0012 = 348
If HR0011 is:

347
or
less
{ CR0021 Off
CR0022 Off
CR0023 On

348
exactly
{ CR0021 On
CR0022 On
CR0023 On

349
or
more
{ CR0021 On
CR0022 Off
CR0023 Off

Figure 16-9
Example G.
Multiple-
Comparison
Program

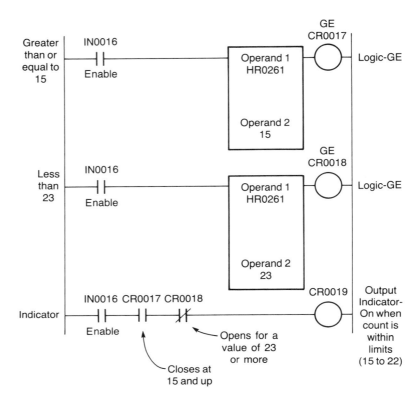

A GE function is used for the lower count. Another GE function is used as an indicator for 22 and is set at 23. IN 0016 enables the functions. Below 15, the top GE function is off, keeping the output in the lower ladder line off.

When the count, starting from 0, reaches 15, the top GE function goes on. The output, CR 0019, turns on when the CR 0017 contact closes. The other contact in the bottom line, CR 0018, remains closed because the lower GE function has not yet come on.

When the up count reaches 22, the output is still on. When the count reaches 23, the lower GE function goes on. Its contact in the lower ladder line opens. The output is therefore turned off at 23 and beyond.

EXERCISES

Construct and test PC COMPARE function programs for the following problems:

1. A light is to come on only if a PC counter has a value of 45 or 78. Hint: two EQ functions with outputs in parallel.

2. A light is to be on if a PC counter does not have values of either 23 or 31.

3. A light is to come on if three input numbers have the same value. Hint: use two functions with the same registers and two contacts controlling an output coil.

4. An output is on if the input count is less than 34 or more than 41.

5. Same as exercise 4, but if the count is 37, the output is on, also.

6. An output is to be on if the count is between 34 and 41. The count includes 34 and 41.

7. Same as exercise 6, but if the count is 37, the output is to be off.

The Skip and Master Control Relay Functions

17

At the end of this chapter, you will be able to

☐ Describe the operation of the SKIP function.
☐ Describe the operation of the MASTER CONTROL RELAY function.
☐ Apply the SK and MCR functions to operational applications.

INTRODUCTION

Both the SKIP (SK) and MASTER CONTROL RELAY (MCR) functions are powerful programming tools. The SK function allows us to skip, or bypass, a chosen portion of a ladder sequence. The coils and functions skipped remain in the state they were in during the last scan before SK was enabled. SKIP is similar to the JUMP command found in many program languages. SKIP enables us to effectively branch to a different portion of the program.

MCR operates similarly. When MCR is enabled on, the ladder diagram functions normally. When MCR is not enabled, a specified number of coils and functions are "frozen" in the off position. Coils in the "frozen" section will then stay off even if their individual enable lines are turned on.

The difference between the two functions is that SKIP leaves the next specified number of ladder lines in their previous on or off state. MCR turns the next specified number of ladder lines to the off state. Another difference is that SKIP is active when enabled and MCR is active when not enabled.

THE SKIP FUNCTION

The SKIP (SK) function, illustrated in figure 17-1, allows a portion of a PC program to be bypassed when its coil is enabled. The enable line of

Figure 17-1
The SKIP
Function

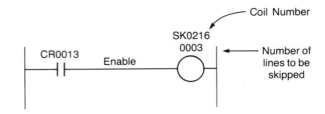

the function is energized when the skip of one or more subsequent lines is desired. In addition to programming a coil number in the usual manner, the number of lines to be skipped is also specified and programmed as shown.

Figure 17-2 shows a basic application of the SKIP function in a program. The eight-line program used for illustration has seven lines with output functions. A SKIP function is included on the third line of the eight. When the SK is off, the other seven functions operate in the normal manner. When the seven lines corresponding to inputs are on, their outputs are on, and when inputs are off, outputs are off. For this illustration, the value for number of lines to be skipped will be set at 3.

Figure 17-2
SKIP Function
PC Operation

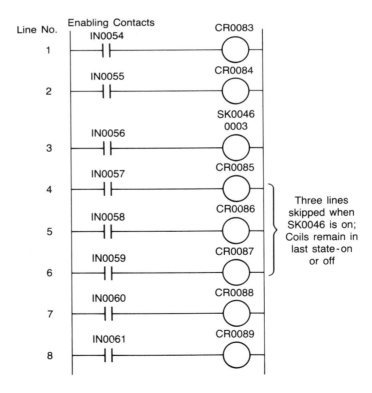

When the SKIP function (set at 3) is turned on, the first two lines will function as usual. However, the next three lines, 4 through 6, will stay on or off in their previous state. With SKIP on, changing the input on-off status feeding the coils on lines 4 through 6 will have no effect on output coils 4 through 6. Coils on lines 4 through 6 will retain their previous states. Lines 7 and 8 will continue to operate normally, unaffected by the SKIP function's operation. Seven and 8 could also be skipped if we had inserted a 5 in place of 3 as the number of lines to be skipped by the function. When SKIP is turned off, the ladder will operate normally.

SKIP Application

There is a production line with eight stations, each of which can perform an assembly operation as the product comes down the line. Depending on the individual part number, all of the eight operations may or may not be set up and carried out. The pattern of whether or not the operations are to be set up and performed is stored in registers, as will be explained in detail in chapter 19. Each of the eight stations is set up to operate or not, according to register bit statuses. A bit of 1 says turn the setup on, and a 0 says turn the setup off.

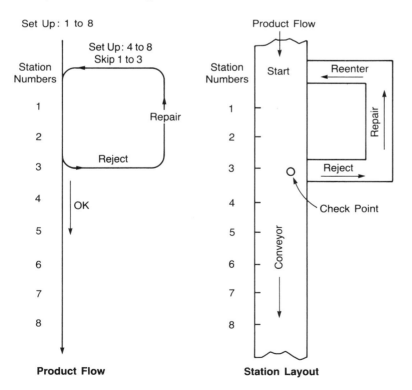

Figure 17-3
SKIP Function
Application
Layout

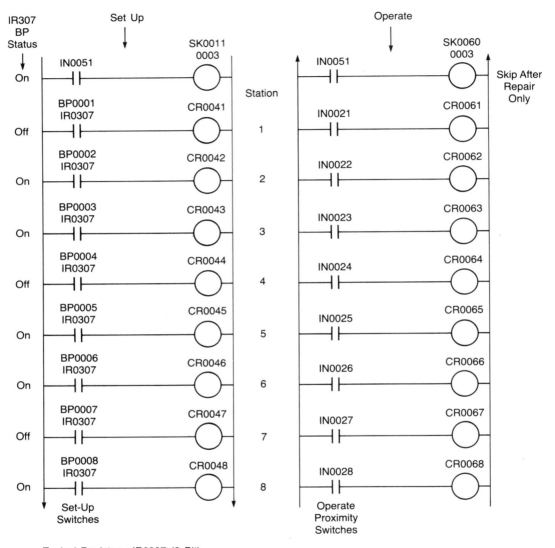

Figure 17-4
SKIP Application Program

At the third station, an inspection takes place. If the part is good, it continues down the operating line; if it is bad, it is shunted to a side conveyor and repaired. After repair, the part reenters at the beginning of the conveyor. The product flow and conveyor layout are shown in figure 17–3.

When a part arrives at the beginning of the line, a sensor (not shown) detects the presence of the parts at the beginning of the conveyor and causes the eight stations to be set up for operation. The sensor causes register contents to turn each of the eight setup switch contacts (the BP/IR contacts) on or off. Figure 17–4 shows this setup system on the left. The setup functions are CR 0041 through CR 0048. As the part proceeds down the conveyor, each operation is performed (if set up) when the part is detected by sensors at each station. These sensors are IN 0021 through IN 0028, as shown on the right of figure 17–4. The operations are CR 0061 through CR 0068.

If a part is rejected at station 3, it is shunted to repair. Later, when the repaired part reenters the conveyor, the setups of stations 1 through 3 do not have to be reset. Unnecessary reset is prevented by the two SK functions, 0011 and 0060. The two SK functions are turned on by a sensor at the repair reentry point.

THE MASTER CONTROL RELAY FUNCTION

The MASTER CONTROL RELAY (MCR) function operation is similar to the SK function. Figure 17–5 shows a typical MCR function. When its enable line is energized, it turns on. When MCR is off, the number of following ladder diagram lines specified are turned off. In contrast to the SK operation where lines were skipped, the MCR turns the following specified number of lines to the off state.

Figure 17–6 shows how the MCR function operates in a program. There are eight lines. The third line is the MCR function. The other seven lines are contact-coil functions. For "fail-safe" reasons, the MCR must be turned on to be inactive. If the function goes off for some reason, it is active and turns the specified lines off, also. When MCR is on, the other seven lines operate normally. When MCR is off, the next three lines, 4 through 6, are turned off. Lines 1, 2, 7, and 8 are unaffected. With MCR off, there is no way to turn on coils 4 through 6 by energizing their enable lines. When MCR is turned off, the ladder operates in the normal manner.

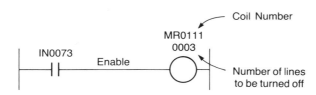

Figure 17–5
The MCR
Function

Figure 17-6
MCR Function
PC Operation

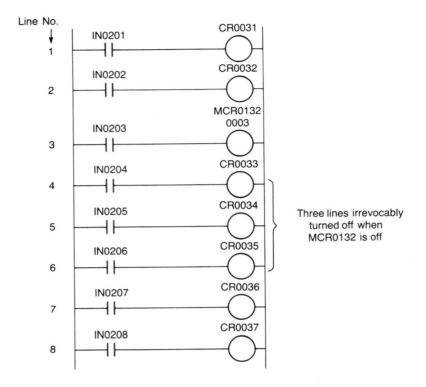

MCR Application

A production line example similar to the SK example will be used for the MCR illustration. There are again eight production stations. Whether each station operates for a given part number as the part goes past depends on the setup (not shown). Each station's operation is initiated by proximity switches at each station. The proximity switches are IN 0081 through IN 0088. Figure 17-7 shows the production line layout and product flow.

Station 5 is an inspection station. Rejected parts are shunted to a repair conveyor. After repair, the part reenters the conveyor. When it reenters, it turns on IN 0011, which turns on and seals an MCR relay, CR 0021. The first five steps are therefore not repeated for the part, because the first five operations are prevented by the MCR. When the part gets to station 5, the MCR is unsealed, enabling stations 6 through 8 to function. These last three steps were not performed the first time through, but are now performed to complete the process. The MCR program for these operations is shown in figure 17-8.

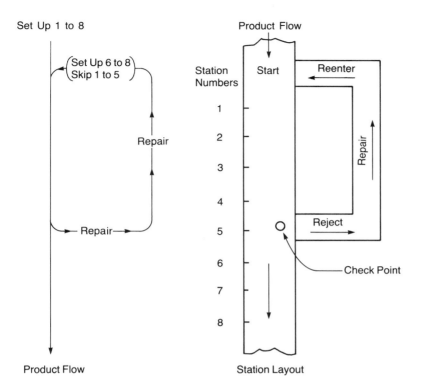

Figure 17-7
MCR Application Layout

Figure 17-8
MCR Application Program

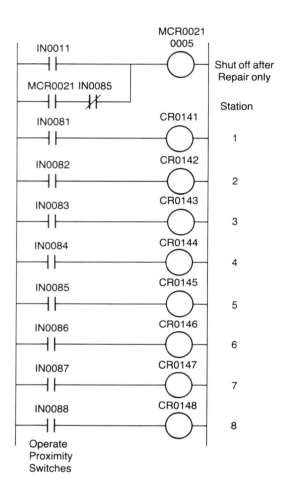

EXERCISES

1. For the 12-ladder line program in figure 17-9, insert three SK functions one at a time. The first problem is to skip lines 3 through 5. The second problem is to skip lines 8 and 9. The third is to skip lines 3 through 11. Program them and check out their operation. When the SK is on, changing input status should not affect the previous status of outputs for the lines to be skipped. As an added problem, insert the line 8 and 9 skip and also the line 3 through 11 skip at the same time. What happens when either one or both are turned on?

2. Repeat exercise 1 using the MCR function instead of the SK functions. When MCR is on, all MCR-designated lines should be off. Additionally,

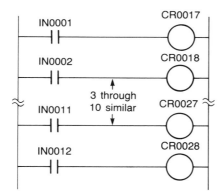

Figure 17–9
Diagram for
Exercise 1

changing any control line's input status for MCR-controlled lines should have no effect on its output.

Repeat the added problem in exercise 1 using MCR instead of SK.

3. Devise an MCR system to control the assembly line shown in figure 17–10. All 15 stations are to function as set up by one of two registers. Short and tall parts are sent down the line. Short parts get all 15 operations, if specified. Lines 6 through 9 are omitted for the tall parts only. Therefore, have operations 6 through 9 turned off by an MCR function when a tall parts goes by. Tall parts are detected by a limit switch just after station 5. After station 9, another limit switch unseals the MCR function so the next part, whether large or small, is again set up for all operations.

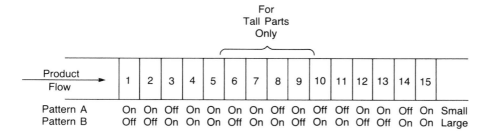

Figure 17–10
Diagram for
Exercise 3

Pattern A is for a product of tall and short parts.
Pattern B is for another product of tall and short parts.

Data Move Systems

18

INTRODUCTION

All computer systems, including PCs, have the ability internally to move data, numbers, and bits from one location to another during computer operation. In smaller PCs, the moving of data from one register to another is carried out automatically internally. The movement of data takes place, but we cannot monitor or control its operation.

In medium and larger PCs, programming functions are available to control data moves. The PC's programmable data moves take data from one place and move it to another. This chapter explains three types of these data-moving programming functions. First, the chapter will cover the basic system of moving one register's contents into another register. The basic MOVE function takes a word, byte, or group bit pattern from one place and moves it to another.

The second type of PC data move involves moving groups of data from two or more consecutive registers to two or more other consecutive registers. This second type is usually designated as a BLOCK MOVE. It moves a consecutive group of register data patterns to another consecutive group of registers.

The third type of data move involves two subtypes. One type sequentially moves data from a designated group of registers into a single register. The other subtype takes the data value from a single register and moves its value (which is normally varying) sequentially to a portion of a table.

In all moves, the contents of the original source register are retained. You essentially then duplicate the source register's value in another register. Conversely, the destination register, which receives the duplicated new data, loses its previous value. In other words, the original value, in the receiving register before the move, is normally lost. If you wish to keep its original value for reference, additional programming is needed to duplicate and store it elsewhere before the move.

THE MOVE FUNCTION

Many PC models denote the data move function as MOVE. Others use a GET and PUT system to accomplish the data moves. We shall use the MOVE system for illustrations. Figure 18–1 shows the elements of a MOVE function. When the function is turned on through the enable circuit, the bit pattern from the specified source register is duplicated in the specified destination register. The source register is unchanged. The destination register pattern is replaced and lost when the new value is brought in. The function coil goes on when the MOVE function is completed. The coil operation can be used to interlock MOVE functions when there is more than one MOVE function in the program. Interlocking prevents energizing two or more contradictory moves when only one is desired. For example, two moves to the same register cannot take place at the same time.

The types of registers that are accessible by moves will vary between PC models. Source locations are generally input registers, output registers, holding registers, or internal registers. Sources also include output and input group registers. Destination locations are the same except input and output register groups are usually not available as destinations.

Move Applications

In chapter 12, the timer program examples all used constant value numbers for time intervals. In PCs, it is possible to use a register's numerical content for the timing interval. For this situation, the numerical value

Figure 18-1
The MOVE
Function

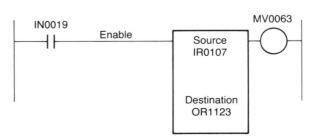

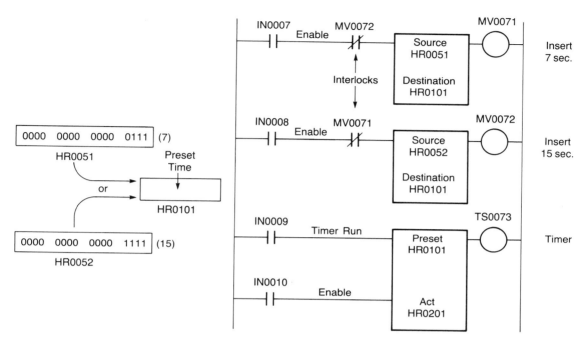

Figure 18-2
MOVE Timing Example

of the holding register is specified as the time interval in the timer functional block.

It is often necessary to quickly change the time interval of a process timer. We can accomplish the change by varying the numerical value of the timer's time-interval holding register with the MOVE function.

Figure 18-2 shows how either of two values of time may be moved into a PC timer. The two different times are 7 and 15 seconds (shown in binary). Either time may be entered into the timer by closing either input switch 45 or 46.

We will give a second example of single register moves, this time using an addition process. Suppose that the PC ADD function can operate only with holding registers as operands. Suppose, also, that the numbers to be added enter the PC into input registers, not holding registers. Furthermore, the result of the addition is to go out of the PC into an output register. PC data moves are therefore required to accomplish the total register configuration.

Figure 18-3 shows a register designation problem in block diagram form. The two numbers to be added enter the PC in input registers. The data must be transferred to holding registers before addition can take place. After the addition is made, the result is located in a holding register.

Figure 18-3
MOVE—
Addition
System

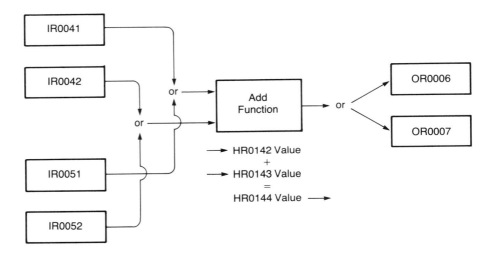

The output of the addition is required to appear in an output register. Another MOVE transfer is then required at the output.

Figure 18–4 illustrates how the PC would be programmed to accomplish the total addition process of this example. IN 0007 closure causes one group of numbers to be placed in the ADD function. IN 0008 moves in other numbers. IN 0010 causes the addition to take place. IN 0017 or 0018 causes the resulting number to be sent to either of two designated outputs.

MOVING LARGE BLOCKS OF DATA

It is sometimes necessary to move more data than the quantity that will fit into one address or register. One option is to use a number of individual MOVE functions. A better solution is to use one PC function that will move many consecutive registers' data at once. These are called BLOCK TRANSFER (BT) functions. Suppose we needed to move 147 bits from one location to another, but have only 16-bit registers available. We would need to use nine full 16-bit registers (144 bits) plus part of another (3 bits). One BLOCK TRANSFER function does the work of ten MOVE functions in this case.

Figure 18–5 illustrates the BLOCK TRANSFER function. In the functional block, specify the number of registers to be moved. Also specify the last register of the input sequence from which the data comes. Some PC systems require the programmer to specify the first register instead of the last; this example will use the last-register system. Finally, specify the last register of the destination register sequence where the data is to be delivered. As usual, the enable input causes the data to be transferred.

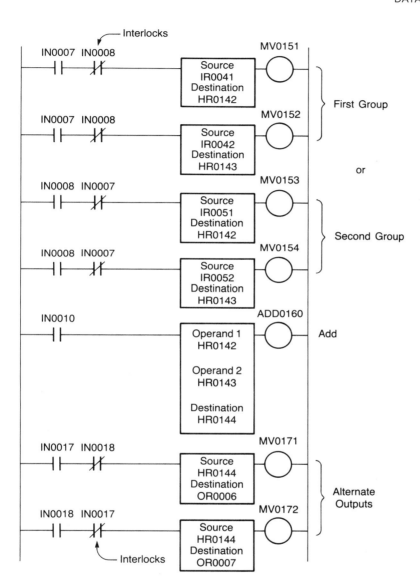

Figure 18-4
MOVE—
Addition
Program

The output coil comes on when the transfer of all registers is complete. The coil operation can be used to verify that the new data pattern in the output registers is complete. BLOCK TRANSFER uses up scan time in proportion to its size; a large amount of computer time can elapse for the BT operation. If the output data were to be utilized in the middle of the transfer, both old data and new data would be in the receiving registers, which could produce some operational problems.

Figure 18-5
BLOCK
TRANSFER
Function

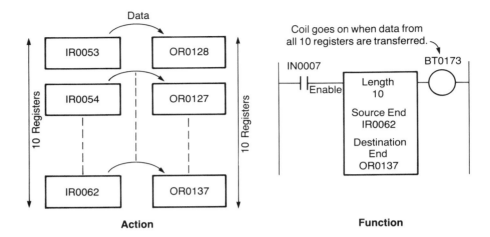

TABLE AND REGISTER MOVES

In the previously described two-move functions, we first moved data from one register to another register. Second we moved data from one consecutive register group to another consecutive register group of equal length. A third type of move involves tables and a single register. The TABLE-TO-REGISTER (TR) function moves data sequentially from a specified portion of a large listing of data to a single register. Conversely, the REGISTER-TO-TABLE (RT) function moves data sequentially out of a single register into a specified portion of a table of registers.

Figure 18-6 shows in block diagram form how the TR function moves data. In a typical application, the receiving register operates a number of machines by bit picking, which will be completely described in chapter 19. As different table register patterns are moved into the receiving register, the machines' on-off patterns will change.

A typical PC function used to accomplish TR moves is shown in figure 18-7. It operates similarly to other MOVE functions. The TR function is first programmed for table length, which is the number of registers to be sequentially inputted. The second line in the TR function is the point at which the table transfer operation is to end. For example, the specified table length is 14 and the specified end register is IR 0058. The table of registers utilized would run from IR 0045 through IR 0058.

The third programmed input to the block is the pointer location, found in many, but not all, PCs' RT and TR functions. The pointer is used to point to the register being moved at any given moment. The pointer location can be used for information only or for control purposes. Finally, a single register destination for the successive data values is specified as a fourth input to the functional block.

The function is enabled when the lower input line is on. When the middle line, reset, is off or turned off, the function is reset to the first register.

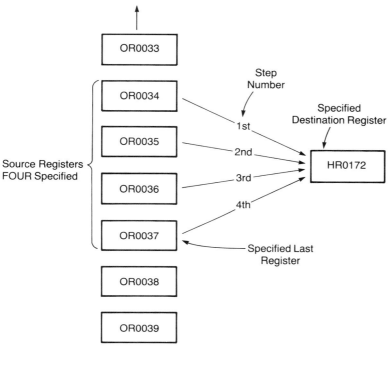

Figure 18–6
TABLE-TO-
REGISTER
MOVE System

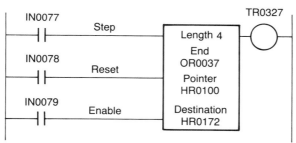

Figure 18-7
The TABLE TO
REGISTER
MOVE Function

The operational pointer, if included in the function, is also set at the first register when reset is off. When reset is on, the function is operational and can be stepped. The top line is the step line. Whenever the step line is turned on, the function transfers data and moves down one register. To step again, the top step line must be turned off and back on. For time interval operation through the table, a timer can be used to do the stepping. A timer contact would be used as the step contact for the function. Chapter 20 will discuss a more advanced form of the TR function, the SEQUENCER function.

The REGISTER-TO-TABLE (RT) function is similar to the TR function. It moves data from a single register sequentially into a specified num-

Figure 18-8
REGISTER-TO-
TABLE MOVE
System

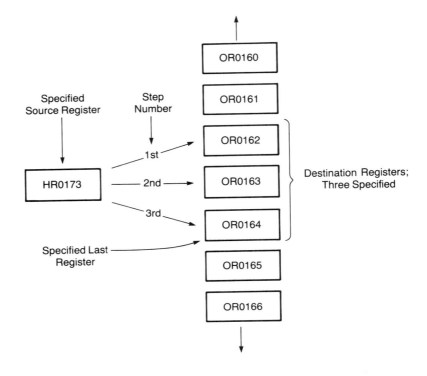

ber of consecutive registers. A block diagram of an RT move is shown
in figure 18-8.

Programming RT moves is similar to programming TR moves. The
input lines operate in the same manner. Table length denotes how many
destination registers are to be used. The table end is the last destination
register to be used. The pointer in the RT function operates similarly to
the pointer in the TR function. The source in the RT function specifies
the one register from which the data is to come.

Figure 18-9
The
REGISTER-TO-
TABLE MOVE
Function

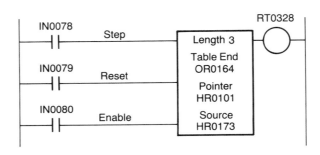

An RT application might be the periodic recording of data. A single register could be programmed to indicate the value of a varying process parameter. The single register's value is constantly changing as the process changes. This register would be used as the source. To record its value every 10 seconds for 5 minutes, we would need 6 times 5, or 30, registers to record the required sequential readings. The function's step line would be pulsed every 5 seconds. The table destination length would need to be 30 registers. The 5-second interval results would then be recorded in order. The 30 sequential values would then appear in order in the specified series of 30 destination registers.

EXERCISES

1. A PC counter is used for the process shown in figure 18–10. During its operation, there are three different count limits to be used. The count chosen depends on the particular product being produced. The three count values are stored in IR 0022, 0023, and 0024. The three limit counts are 5, 12, and 17, respectively. Inputs 0034, 0035, and 0036 are used to insert the three respective limit counts one at a time. The counter uses HR 0345 for the preset time limit value. Design a PC ladder program to accomplish the change in the process count limit using one counter and three MOVE functions.

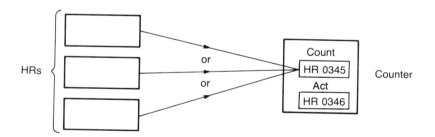

Figure 18–10
Diagram for
Exercise 1

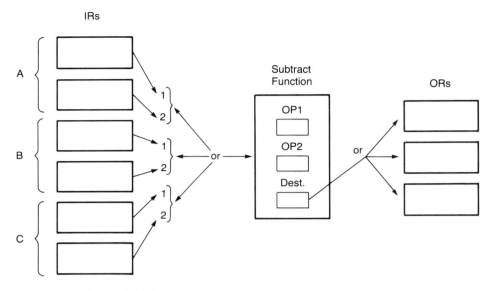

Figure 18-11
Diagram for Exercise 2

Figure 18-12
Diagram for
Exercise 3

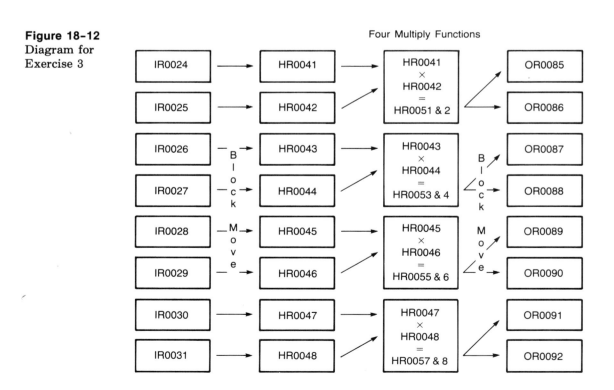

2. A PC subsystem has a SUBTRACT function, as shown in the block diagram in figure 18–11. The SUBTRACT function has three possible sets of two input numbers. The resulting SUBTRACT output is to be moved to one of three output registers. The output register chosen to receive the SUBTRACT result is determined by the position of a selector switch. Design a PC ladder program to carry this out using six input MOVE functions and three output MOVE functions along with the SUBTRACT function.

3. Use two BLOCK MOVE functions to accomplish the process shown in the block diagram in figure 18–12. The first BLOCK MOVE function moves the 8 input registers' contents to the operand 1 and 2 positions in the four MULTIPLY functions. The resulting MULTIPLY values, which are two registers wide, are then to be moved by a second BLOCK MOVE function to 8 output registers as shown.

4. There are 30 bit patterns of 27 bits to be moved sequentially into OR 0011, one every 7 seconds. Design a double TR function program with a timer to accomplish the data transfer. Two TR functions are required because there are more than 16 bits to be transferred, the amount available in one register.

5. You are assigned the task of recording lap times at the Indianapolis 500. There are 200 laps. A laser sensor determines the exact moment your assigned car passes the start/finish line and turns on IN 0042. You have 200 registers in which to place the time at each lap. Design a TR function system to accomplish the recording.

6. Develop a time system for exercise 5 that will also record the elapsed time for each lap. Exercise 5 recorded total time at each lap's end. Add appropriate PC circuitry with a SUBTRACT function and a second set of recording registers for recording the individual lap times.

SECTION FIVE

FUNCTIONS INVOLVING INDIVIDUAL REGISTER BITS

Utilizing Digital Bits

19

At the end of this chapter, you will be able to

☐ Describe the PC digital bit control system.
☐ Describe the BIT PICK CONTACT function and its use.
☐ Use digital bits to turn outputs on and off.
☐ Modify and control digital bits in a register.
☐ Use shift registers to move digital bits within and through registers.
☐ Apply digital bit register systems to process control programs.

INTRODUCTION

Medium and large PCs have capabilities to work with digital bits. Instead of controlling output devices from individual contacts, these PCs use register bits in groups. For example, if the on-off of 16 machines must be controlled, just one of the 16 bits in a 16-bit register could control each of the 16 machines. If there are 157 machines to turn on and off, only 10 of these 16-bit registers are needed for their on-off control ($157/16 = 9.815$, or 9 registers plus part of a 10th one). By contrast, a contact-coil ladder control would need 157 ladder lines in the program.

The PC not only uses a fixed pattern of register bits, but can easily manipulate and change individual bits. The PC can pick, set, latch, and manipulate the individual bits in chosen registers. It also can shift the register contents to the right or left. Register shifts can be set to move the bits one position per input pulse. Shifts may also be set for multiple position movement (2, 3, or more). This MULTIPLE BIT SHIFT function is often designated the N-BIT Shift.

Functions discussed in other chapters also play a part in process control with digital bits. For example, MOVE enables you to replace the entire register contents in order to change the 16 output commands. If you want an on-off pattern changed, shift in an appropriate new register pattern. Moves of data into registers can be done for one register only, but data moves can be made for a consecutive series of many registers.

The digital bit system is the foundation of multiple machine control. The bit system is used extensively in all types of automation systems. One very powerful bit control system is the drum controller/sequencer to be discussed in chapter 20.

BIT PATTERNS IN A REGISTER

In some PCs, the internal slots for memory and operation are called addresses. In others, the slots are called registers. This chapter will again use the word *register*.

Chapters 11 and 12 dealt with counters and timers, which require an awareness of register use and contents. This chapter is not concerned with a register's numerical value, but only its binary bit pattern status; that is, its pattern of 1's and 0's.

For example, start with the register bit setting shown in figure 19–1. A bit pattern has been inserted into the register by calling it up on the screen and then keyboard-inserting its desired bit values. For illustration purposes, the register bits in HR 0207 have been arbitrarily given the values shown. HR 0207 now has an equivalent BCD value of 7851 and an equivalent decimal value of 30,801. These BCD and decimal values are irrelevant at this point; only the binary pattern shown is useful. Binary bit patterns can be applied to (and from) any type of register, not just holding registers.

Figure 19–1
Register with a
Binary Value

0111 1000 0101 0001   HR 0207

THE BIT PICK CONTACT

Suppose you wish to have outputs CR 0081 and 0082 controlled by a register bit status. To have CR 0081 controlled by bit 11 in HR 0207 and CR 0082 controlled by bit 12, you would designate the contacts as shown in figure 19–2. A menu appears when you press the contact key on the keyboard. Instead of choosing CR or IN, as you have been doing, in this case you would choose BP.

Take the first 10 bits (from the right is standard) and use them to control 10 outputs, as shown in figure 19–3. The outputs with a feeder bit

Figure 19–2
BIT-PICK CON-
TACT Control

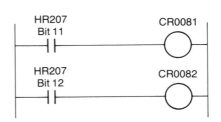

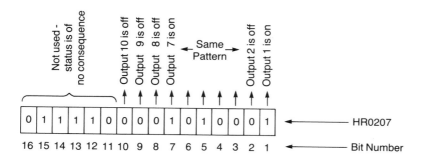

Figure 19-3
Ten Outputs
Controlled by
Ten Register Bits

of 1 would be on, and those with 0 bit would be off. If you modify HR 0207 to another pattern of bits, the outputs would change status accordingly. An appropriate BP contact system would be used.

CHANGING A REGISTER BIT STATUS

Suppose you wish to change bit 4 in HR 0207 from 0 to 1. Call up register HR 0207 on the screen and completely rewrite its bit pattern. Pushing *Return* would insert the pattern into the PC CPU. Otherwise, move the cursor over bit 4 and change bit 4 only. This change process is very slow.

Bit status changes are more quickly accomplished by using one of three PC functions. These are BIT SET (BS), BIT CLEAR (BC), and BIT FOLLOW (BF). We will illustrate the first function using the 4th bit of holding register HR 0207. When the BS function is enabled in figure 19-4, bit 4 of HR 0207 is set to 1 (if it was not already a 1). Turning the function off would have no further effect on the bit—it would remain a 1.

The BIT CLEAR function, shown in figure 19-5, has the opposite effect of BIT SET. The example in figure 19-5 operates on bit 5 of HR 0207. When enabled, the BC function would change bit 5 from 1 to 0. If you had applied BC to bit 6, nothing would happen, because bit 6 is already a 0. When BC is turned off, nothing further happens.

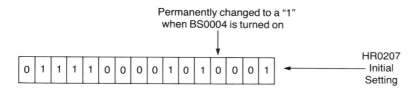

Figure 19-4
The BIT SET
Function

Figure 19-5
The BIT CLEAR
Function

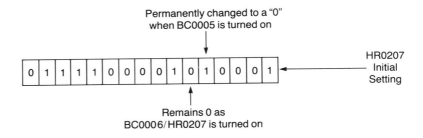

Figure 19-6
The BIT FOL-
LOW Function

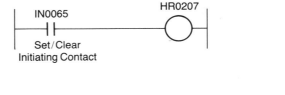

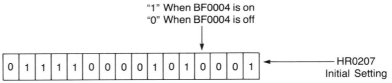

There is one more bit-operating function, the BF, or BIT FOLLOW, function. Go back to bit 4 of HR 0207. Figure 19–6 shows the BF function as applied to this bit. When enabled, the function sets the bit to 1. When disabled, or off, the function sets the bit to a 0. Notice how BF differs from BS and BC: on and off are both active and significant in the BIT FOLLOW function.

APPLICATION OF BS, BC, AND BF

Figure 19–7 shows a board-painting process that uses the bit modification functions. White, square boards are to be painted red in certain areas. There are 16 square sections on the board, as shown. There are 16 spray guns, one above each of the 16 sections, that spray perfect squares through a template.

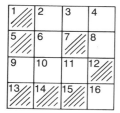

White Board Portions
Painted Red ⟶ ////

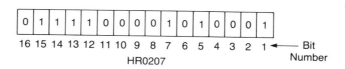

Figure 19-7
Spray-Paint Pat-
tern and Program

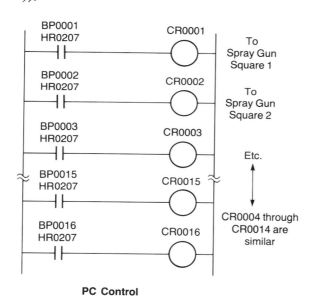

PC Control

When the spray guns (red) operate in the pattern of the original HR 0207, the red/white pattern will be as shown. A bit of 1 would cause the corresponding spray to take place. Each spray gun's operation is controlled by a corresponding PC output coil, which is controlled by a PC contact as shown. The input contact is described by two specified lines: a register number line and a bit number line. Note that you could use an output group register instead of 16 output coils. This illustration uses individual output coils.

As board model patterns are changed throughout the day, the red/white patterns will change. For example, square 15 is to be changed from red to white. This change may be made (permanently) by applying the BC function as shown in figure 19–8. To change square 3 from white to red, apply a BS function to bit 3. To change square 10 back and forth repeatedly between red and white, apply a BF function as shown in the figure.

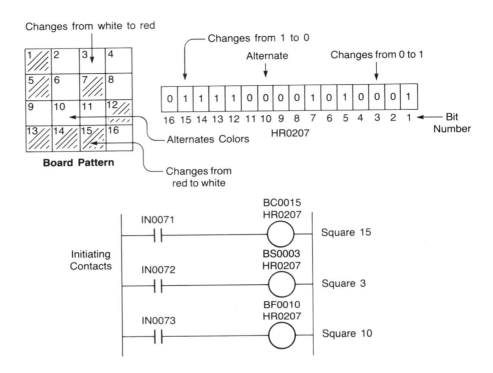

Figure 19-8
Revised Spray-
Paint Pattern
and Program

A word of caution: using BF on the same bit as BS or BC can cause problems; for example, if both BC and BF were working on bit 7 of HR 0207, the BF would probably override the BC in most PC models.

SHIFT REGISTER FUNCTIONS

The SHIFT REGISTER function enables the operator to move digital bits within and through the PC registers. This is accomplished through the use of the SHIFT RIGHT, SHIFT LEFT, ROTATE, and MULTIPLE SHIFT functions. This section discusses each of those functions.

Shift Right

Figure 19–9 shows the operation of a SHIFT RIGHT (SR) function. This explanation uses only one register. Later sections show how to use more than one. There are normally three inputs to the functional block: the bottom input is normally the enabling input, as in previous PC functions; the middle input determines whether a 1 or a 0 is inserted into the register when shifting; and when the top input is activated, the register shifts all bits one position to the right and a new bit is added on the left.

Whether the bit in the vacated register on the left becomes a 1 or a 0 depends on whether the middle input was on or off when the shift took

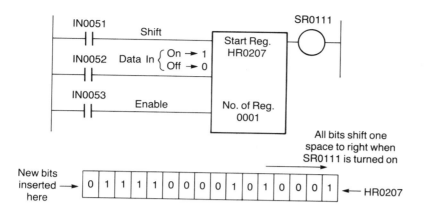

Figure 19-9
Shift-Right
Function—One
Register

place. If the middle data line is on, a 1 is entered and, for off, a 0. The pattern in figure 19-10 illustrates how to use a shift register to produce the original values in HR 0207 shown in figure 19-1. A shift is made 16 times. This illustration starts with all 0's in the register. The register might have had any other pattern; it would not affect the final pattern, since all previous bit 1's and 0's are pushed out after 16 steps.

Another important part of this SR function is that the coil, or output, status follows the status of the bit on the right. A 1 produces an output on, and 0 results in output off. A later application explains how output on-off is used. The bits are normally lost when they are pushed off to the right; however, the bit status can be saved and reused in the rotate function, which will be discussed shortly.

Now suppose we need to control 45 machines or functions. Sixteen bits would not be enough to control the process; we would need to shift through three registers to cover the 45 outputs by placing the number 3 in the function block that asks for the number of registers. If we put HR 0207 in as the starting register, we will shift through HR 0207, 0208 and 0209. Multiple register shifting is shown in figure 19-11.

Shift Left

SHIFT LEFT (SL) functions operate exactly like SHIFT RIGHT, except that bit status is inserted on the right. The bits shift to the left and leave on the left. The output coil status normally then follows the status of the last bit on the left.

Rotate

Suppose you wish to save the bit sequence status that leaves and is lost when using shift registers. You also may want to repeat a pattern again and again. This may be accomplished by using the REGISTER ROTATE

Figure 19-10
Operation of the
SR Function

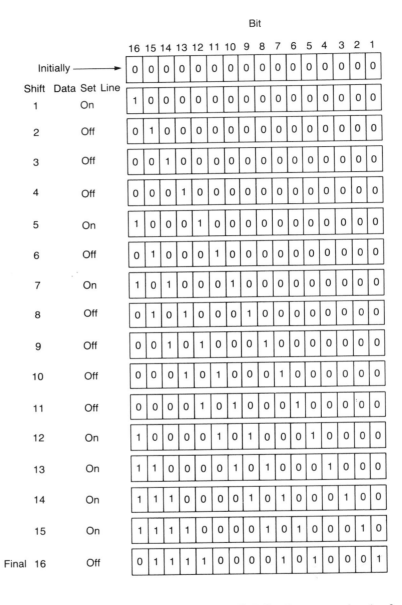

Shift	Data Set Line	16	15	14	13	12	11	10	9	8	7	6	5	4	3	2	1
Initially →		0	0	0	0	0	0	0	0	0	0	0	0	0	0	0	0
1	On	1	0	0	0	0	0	0	0	0	0	0	0	0	0	0	0
2	Off	0	1	0	0	0	0	0	0	0	0	0	0	0	0	0	0
3	Off	0	0	1	0	0	0	0	0	0	0	0	0	0	0	0	0
4	Off	0	0	0	1	0	0	0	0	0	0	0	0	0	0	0	0
5	On	1	0	0	0	1	0	0	0	0	0	0	0	0	0	0	0
6	Off	0	1	0	0	0	1	0	0	0	0	0	0	0	0	0	0
7	On	1	0	1	0	0	0	1	0	0	0	0	0	0	0	0	0
8	Off	0	1	0	1	0	0	0	1	0	0	0	0	0	0	0	0
9	Off	0	0	1	0	1	0	0	0	1	0	0	0	0	0	0	0
10	Off	0	0	0	1	0	1	0	0	0	1	0	0	0	0	0	0
11	Off	0	0	0	0	1	0	1	0	0	0	1	0	0	0	0	0
12	On	1	0	0	0	0	1	0	1	0	0	0	1	0	0	0	0
13	On	1	1	0	0	0	0	1	0	1	0	0	0	1	0	0	0
14	On	1	1	1	0	0	0	0	1	0	1	0	0	0	1	0	0
15	On	1	1	1	1	0	0	0	0	1	0	1	0	0	0	1	0
Final 16	Off	0	1	1	1	1	0	0	0	0	1	0	1	0	0	0	1

function, which is found in some, but not all, PCs. Its operation is shown in figure 19-12. This example uses the same pattern as the 45 functions in figure 19-11. Now the pattern movement is repeated again and again as a result of the rotate automatic reentry system. For the previous shift registers, the pattern would have to be reentered manually or by MOVE for each time through; the ROTATE functions are automatically repetitive.

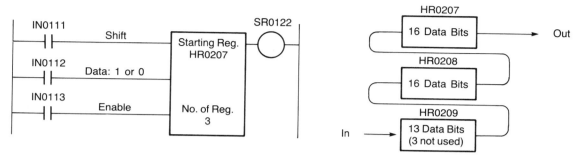

Figure 19-11
Shift-Right Register—Multiple Registers

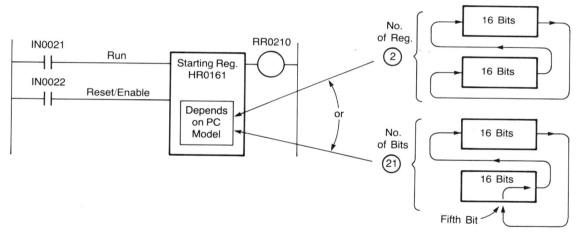

Figure 19-12
The REGISTER ROTATE Function

ROTATE functions may be Rotate Right (RR) or Rotate Left (RL). ROTATE systems can be of two other general types: full-register reentry or partial-register reentry. The full reentry system reentry point can only be at the beginning of a register. With 16 bit registers, you must shift through 32, 48, 64, and other 16-multiples of bits. With the partial reentry type, you may choose the exact number of bits needed. For example, if you need only 27 bits, you could reenter the initial register at the 27-minus-16 point, which is the 11th bit. The point would be specified by entering 27 bits into the block function.

Multiple-Shift Registers

Some advanced-function PCs have SHIFT RIGHT and SHIFT LEFT functions that shift more than one bit at a time. These might be labeled

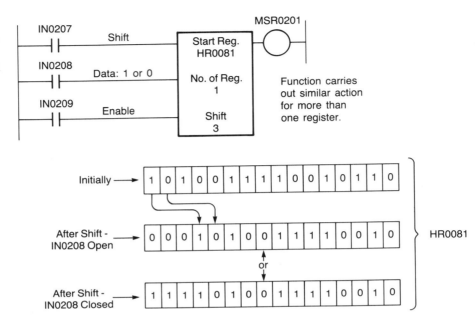

Figure 19-13
Typical MULTI-
PLE SHIFT
RIGHT Function

MULTIPLE SHIFT RIGHT (MSR) and MULTIPLE SHIFT LEFT

MULTIPLE SHIFT RIGHT (MSR) and MULTIPLE SHIFT LEFT (MSL), or N-BIT right and N-BIT left. Figure 19–13 shows a typical multiple-shift register. The MSR or MSL functions need one more piece of input information than the SR and SL functions: a specification of the number of shifts to be made at a time, N.

For example, assume that the number of shift steps, N, is set at 3. For an MSR function, it will put in three 1's or three 0's, depending on whether the serial-in switch is closed or open, respectively. The before and after register pattern is shown in figure 19–12, starting with the original register, HR 0207.

SUMMARY OF SHIFT REGISTER OPERATION

Figure 19–14 gives a summary of the operation of eight types of shift registers discussed in this chapter. The bit identification numbering system in all of these can be of two types. For two registers, HR 0207 and 0208, the total bit numbers can go from 1 through 32; for three registers, it would be 1 through 48, etc. The numbering system can also be 1 through 16 for each register only. See your PC operational manual to determine your system's bit identification scheme.

SHIFT REGISTER APPLICATION—LIGHT PATTERN

Figure 19–15 shows an arrangement of lights to give a flashing, moving arrow pattern that moves to the right. Each light is connected to a PC

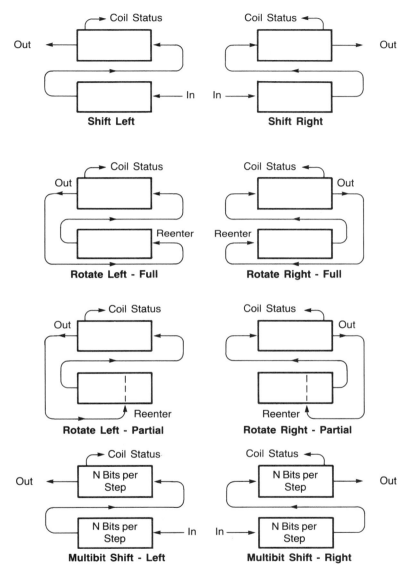

Figure 19-14
Shift Register
Operation
Summary

input terminal. Each terminal is controlled by a bit location in the two registers shown. As the bit patterns are moved to the right, one step at a time, the light pattern moves to the right. The seven 1's in the registers are progressively moved to the right and then up to the next register. 0's are initially entered into the emptied slots. As the bits move, the lighted lamp pattern progressively moves to the right. In this system, a complete arrow (example 1 through 7) does not move all at once (to 8 through 14); the lighted lamps progress to the next arrow, one at a time, starting at the top (1, then 2, etc.).

Figure 19-15
Flashing Arrow
Move Pattern
and Registers

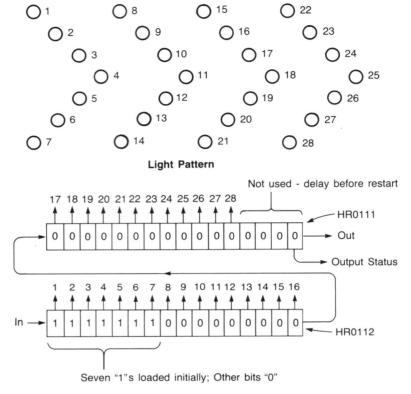

Figure 19-16 illustrates the complete PC function that accomplishes the movement. The speed of movement is controlled by the time set in TT 0151, 5/10 of a second in this example. The SR register is therefore pulsed by the input shift line every 5/10 of a second by the timer. Enabling of the timer and the SR is accomplished by IN 0050. As the 1's move through the registers, the corresponding outputs are progressively turned on by the 28 BIT-PICK functions shown at the bottom. The arrow then moves to the right.

To repeat the process, you could reload the 0's and 1's manually by opening or closing the data line input, or you could use the MOVE system to reload the two operating registers from two master registers. Another possible reload system is to reload from the output status. When the 1's reach the end of the registers, SR 0150 coil goes on for each step. This will cause a 1 to be reinserted into the initial slot. For a 0, the output is off and a 0 will be reloaded. At the end of 31 steps, we are reloaded as shown.

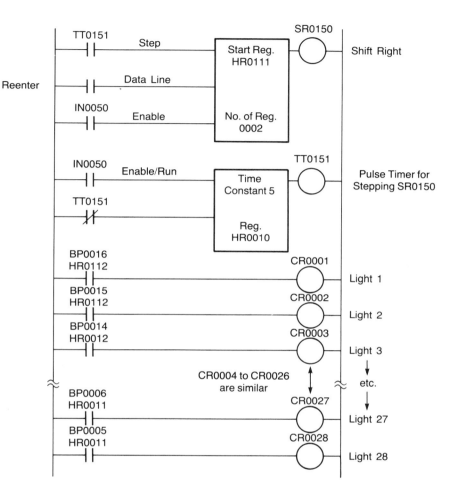

Figure 19-16
PC Program for
Flashing Arrow
Movement

If you had used a ROTATE RIGHT function the reloading would have been 32 automatic steps. An alternate output scheme for figure 19-16 could be one using output group registers.

SHIFT REGISTER APPLICATION—CODE OUTPUT

The second example of the use of the SR (or SL) register uses the output coil instead of the individual bits for control and indicating. It uses the output as a Morse code indicator to a light or a buzzer. In the Morse code system, each letter is assigned a dot-dash configuration as shown in figure 19-17. Follow the Morse code rules to put dots, dashes, and intervals in registers. A dot is 1 on-bit wide (1) and a dash is 3 on-bits wide (111). Appropriate spacing is added between letters and words.

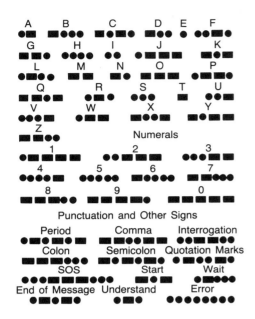

Morse Code System

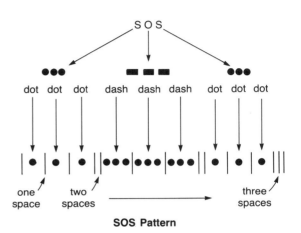

SOS Pattern

Morse Code Rules

1. Dash (−) is three times as long as dot (•)
2. One space between dots and dashes
3. Two spaces between letters
4. Three spaces between words
5. Four spaces between sentences

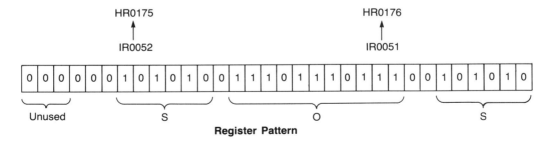

Register Pattern

Figure 19–17
Morse Code in SR Register Form

The resulting pattern for the international distress symbols SOS (save our ship) is put in input registers IR 0052 and IR 0051 in the figure. The bit pattern is then moved into two HRs, 0175 and 0176. The register bits are then shifted to the right to the output. The resulting output on-off pattern represents the code pattern.

Figure 19–18 shows the PC program to produce the coded output. Turning on IN 0010 will move the complete coded pattern from the IR registers into the two HR registers. Then, turning on IN 0012 enables the SR function and starts the pulsing timer. The timer contact in the SR shift line causes a bit to be inserted into the SR every 0.5 seconds. Since the timer is set at 5/10 of a second, a dot (one bit) will be 0.5 seconds long, and a dash (three bits), 1.5 seconds long. Appropriate intervals are added between dots and dashes. To speed up or slow down the rate of code output, the timer's time interval may be changed.

Once the 32 bits have been shifted in, the message is over. We then have on or off signals for the output, depending on the SR data line set-

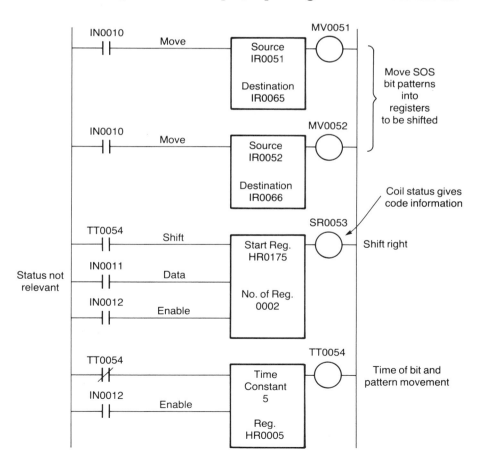

Figure 19–18
Morse Code
Shift Register
Program

ting during shifting. For SOS, we must reload the HR patterns into the IR registers again and start over. An alternate to the reload procedure would be programming in a ROTATE-RIGHT register system, which reloads continuously.

EXERCISES

The first three chapter exercises are applications for register bit control programming. Include a sketch of the hardware needed for each exercise, as well as one for the program.

1. A product moves continuously down an assembly line that has 15 stations. Set up a register-controlled production line for the pattern for product A only. The A pattern is shown in figure 19–19.

Figure 19-19
Diagram for
Exercises 1 and 2

Product	Station Status														
	1	2	3	4	5	6	7	8	9	10	11	12	13	14	15
A	On	Off	Off	On	On	On	Off	On	Off	On	Off	On	Off	On	Off
B	On	On	Off	Off	On	Off	Off	Off	On	On	Off	On	On	Off	On
C	On	Off	On	On	On	Off	On	On	Off	On	On	On	Off	On	Off

2. Next, three products are sent down a 15-station production line. They are product A of exercise 1, and products B and C. The schedule of which stations are on or off for each of the three products is shown in figure 19–19. A selector switch is set according to which product is going down the production line. Design a PC circuit to produce the three products to the required patterns.

3. There are two production lines. Either line may produce products A, B, or C. Expand your program from exercises 1 and 2 to control the two lines.

The final three exercises are applications of shift registers. Again, include a hardware sketch and a PC program for the solution.

4. There is one mini-assembly station that produces two models of watches, E and F. Sixteen operations take place sequentially at the station. Stations are active or inactive, depending on the model being produced. The station pattern is shown in Figure 19–20. At the end of the day, all watches that have not passed the test are disassembled by running the watch assembly operations in reverse. Design a PC system with SR for assembly and SL for disassembly. Use the same pattern register for both the SR and SL functions.

	Station Status															
	1	2	3	4	5	6	7	8	9	10	11	12	13	14	15	16
Pattern E	On	Off	On	On	On	On	Off	Off	On	On	On	Off	Off	On	Off	On
Pattern F	On	On	Off	On	On	Off	On	On	Off	On	On	On	On	Off	Off	Off

Figure 19-20
Diagram for Exercise 4

5. The lights in a circle are 8 degrees apart. The pattern of lighted lights is to rotate, as shown in figure 19–21, in either direction. The 8-degree steps take place every 3 seconds for clockwise rotation. When the pattern is rotating counterclockwise, steps are at 1-second intervals. Design a shift register system to accomplish this rotation.

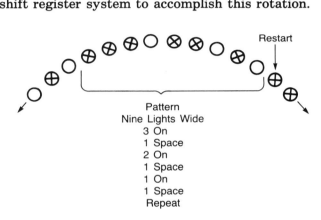

Figure 19-21
Diagram for Exercise 5

6. Design a PC system to put out the word ACME in Morse code to a buzzer. Select a speed of either 4/10 or 7/10 of a second per step. Additionally, select any other word of your choice for the output, and reprogram accordingly.

The Sequencer Function

20

At the end of this chapter, you will be able to

☐ Compare the advantages and disadvantages of a conventional drum switch with the PC DRUM CONTROLLER/SEQUENCER function.
☐ Describe the program layout of a PC SEQUENCER function.
☐ Construct a matrix for a process—output state versus interval and time.
☐ Program a SEQUENCER function into the PC.
☐ Insert time intervals into PC SEQUENCER programs.

INTRODUCTION

The PC SEQUENCER function is often alternately called the DRUM CONTROLLER function. We shall use the function designation DR, instead of SQ, which has already been designated for square root. The SEQUENCER concept has evolved from the mechanical drum switch, which is an important control device, but the PC SEQUENCER function handles large sequencing control problems more easily than does the drum switch. Another advantage of the PC is that its SEQUENCER programming is relatively straightforward and user friendly.

Traditional drum switches are manually operated. If a timing of the steps being controlled by the drum switch is required, manual operation timed by a clock is needed. The PC SEQUENCERS can operate between steps by programmed time sequences. This chapter explains how the PC SEQUENCER function operates and can be applied to control problems.

THE MECHANICAL DRUM CONTROLLER

Figure 20–1 shows a small, electromechanical drum controller. It is a three-position, six-electrical-terminal device. Its electrical internal connections are illustrated in figure 20–2 for each of its three positions. How is this drum switch used in process control? Four motor-reversing applications are illustrated in figure 20–3. Motor reversing is accomplished by revers-

Figure 20-1
Electromechanical Drum Switch

Figure 20-2
Internal Contact Switching for Figure 20-1 Drum Switch

Left	Up	Right
	Handle End	
Forward	Off	Reverse
1 o——o 2	1 o o 2	1 o o 2
3 o——o 4	3 o o 4	3 o o 4
5 o——o 6	5 o o 6	5 o——o 6
	Internal Switching	

Figure 20-3
Motor-Reversing Applications for Figure 20-1 Drum Switch

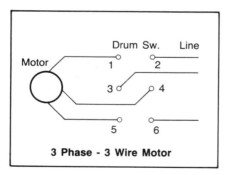

3 Phase - 3 Wire Motor

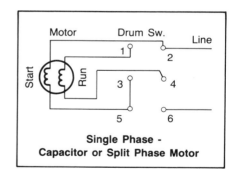

**Single Phase -
Capacitor or Split Phase Motor**

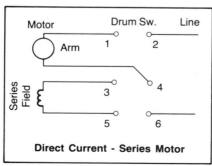

Direct Current - Series Motor

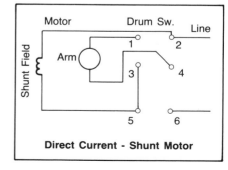

Direct Current - Shunt Motor

228

ing any two leads for three-phase AC, reversing the start leads with respect to the main leads for single-phase AC, or reversing the field leads with respect to the armature leads for DC.

Drum switches are limited to a maximum of 7 positions and about 12 pairs of contacts. The switches cannot handle a process with 27 devices and 138 steps, for example, unlike the PC SEQUENCER function which can easily handle the 27-by-138 control, and more. The electromechanical drum switch in figure 20-1, however, has one major advantage: it is a good, economical control device for handling applications with a fixed sequence and a limited number of required contacts.

A THREE-BY-FIVE SEQUENCE EXAMPLE

To begin this discussion of sequencers, a simple sequence of operation is shown in figure 20-4. Three lights are to be in the on or off state in five consecutive, different conbinations. The five steps are to be in a given sequence, 1 through 5. A 1 indicates the light is to be on and a 0 off.

If you had only toggle switches to run the operation, figure 20-5 shows how to hook them up. Each toggle switch has three electrical poles. By connecting the toggle wires as shown, each toggle would create its required light pattern. Note that only one toggle switch should be on at a time. In this system, one toggle is turned on and then off in sequence.

You could use a PC contact and coil system, as in chapter 11, to produce these same patterns. A PC circuit to do so is illustrated in figure 20-6. In this case, five single-pole switches would feed the input module.

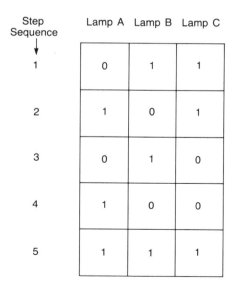

Step Sequence	Lamp A	Lamp B	Lamp C
1	0	1	1
2	1	0	1
3	0	1	0
4	1	0	0
5	1	1	1

Key
1 → On
0 → Off

Figure 20-4
Light Pattern
Sequence

Figure 20-5
Toggle Switch
Pattern Control

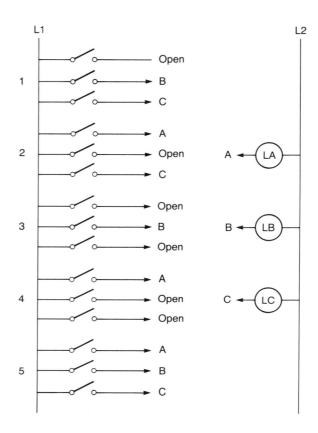

Wiring for the PC system is simpler, but PC programming is time consuming.

Figure 20-7 shows in simplified form how this pattern sequence would be entered into a PC SEQUENCER function. The DR function would be called up on the screen and programmed, and the desired light pattern would be entered into five consecutive registers. The on-off pattern would be programmed as the first three bits of the registers. Then, repeatedly energizing the SEQUENCER function would cause it to step through the registers one at a time. The on-off patterns are then sequentially fed to the first three outputs of a specified output module. The three output lights are connected to these first three output module terminals, which in turn control the output pattern.

THE PC SEQUENCER FUNCTION

Suppose that you had 43 functions to step through 47 steps. Using the previous 3-by-5 example, toggles or coil/contact programming would be very ponderous. However, programming 43 functions by 47 steps is easi-

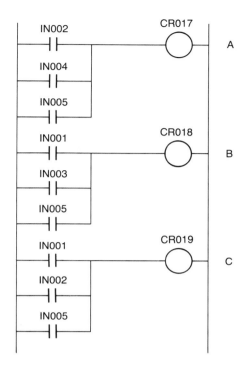

Figure 20-6
PC Sequence Contact/Coil Program

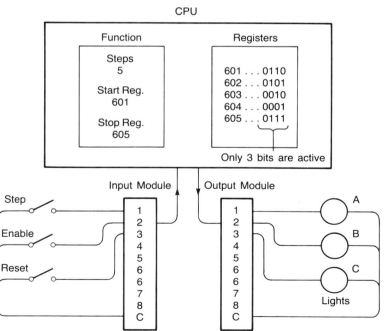

Figure 20-7
PC SEQUENCER Function and Pattern for the Light Sequence

ly accomplished by the PC SEQUENCER function. This section illustrates the advantage of using the PC SEQUENCER by applying it to a 6-by-11 dishwasher control problem. Assume that the dishwasher has six functions to be turned off and on periodically (an actual dishwasher has more). These are:

☐ soap release solenoid
☐ input valve for hot water
☐ wash-impeller operation
☐ drain water valve
☐ drain pump motor
☐ heat element for drying cycle

The operational pattern for the dishwasher is shown in figure 20–8. There are 11 steps, so you need 11 registers to control the pattern. On the right of the figure the required register pattern for correct dishwasher sequencing is shown. The bit patterns are programmed to match the on-

Figure 20–8
Dishwasher
Function Matrix
and Register
Patterns

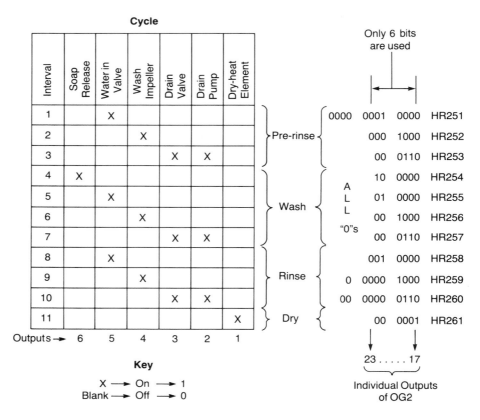

off pattern. Only the first 6 bits of the 16 in the register are active in this program. Since you are using a 16-bit register, the settings of the other 10 bits are irrelevant. These other 10 do not go to an active output.

The problem of time interval must now be considered. Some manufacturers' formats put the times in a group of registers separate from the pattern registers. For these separate registers, two sequencers operate simultaneously. This puts patterns and times together for each step of the sequence. Other PC formats put patterns and times for each step in one register. Either system enables us to have individual times for each step.

The pattern registers are shown on the right side of figure 20–8. A typical PC function that automatically scans through the registers is shown in figure 20–9. The three inputs required for function operation are described by notes on the diagram. The function block requires that four pieces of information be defined during programming:

1. How long is the register string or number of steps?
2. What is the starting register's number?
3. An optional register or address is the location of a step pointer. The step pointer register will indicate which step the program is on during process operation. The pointer-indicated step number is often useful information, especially in a long program.
4. Where does the output on-off pattern go? Output register groups are specified. A 16-bit register needs an output group of up to 16 terminals. This example uses only 6 of the 16. Which terminal group is active is determined by the OG number, as previously discussed in chapter 8. If we choose OG 0002, outputs 17 through 32 are active, and 17 through 22 are used.

In figure 20–9 you must close the step circuit manually to increment the program from one register pattern to the next. When the sequencer

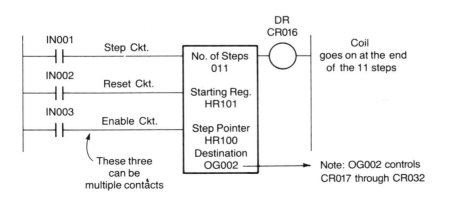

Figure 20-9 Typical PC SEQUENCER Function

(DR) reaches the last step, it automatically restarts at the first step whenever the step line is energized.

Alternately, a pulse timer can be used to step the function regularly as shown in figure 20–10. With this added programming, the function will step every 4 seconds. Additionally, there is a start-stop system to enable the sequence to stop at the end of the 11 steps. Start must be depressed to restart the sequencer at the beginning again. Note that 12 steps are programmed to complete 11 process steps. This is because the sequence stops by unsealing CR 0039 when the last step is started. If you stop at the 11th step, the unsealing never takes place. Stop takes place at the instant you get to step 12. The ladder sequence of figure 20–10 is:

1. Push start (momentary button), CR 0039 on and seals.
2. Sequencer and timer enabled. Process on step 1.
3. After 4 seconds, timer pulses on/off, stepping DR.
4. Repeat for steps 2 through 11.

Figure 20–10
Sequencer with
Timer Pulsing

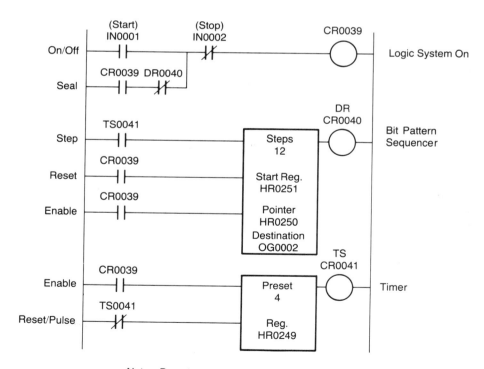

Notes: Do not use any
register for more than one purpose.
Select coil and function numbers
that are not used elsewhere, for example,
as one of the OG series.

5. When DR steps to step 12, DR comes on. CR 0039 unsealed.

6. Ladder is reset and ready for step 1.

7. Anytime stop is depressed, the system stops and resets. (This may or may not be safe. Other programming may be necessary to stop the process at the step it is on.)

You may also want each step to have different times. If the times are all multiples of 4 seconds, you can use repeat registers and lengthen the program. Assume that the steps have time interval values of 8, 4, 12, and so on, as shown in figure 20–11. The program could be lengthened as shown in the figure to accommodate these different times. The PC program for this multiple interval is shown in figure 20–12.

If you want variable times instead of multiples, you would use two sequencers operating together. One sequencer would step through one group of registers for the output patterns and the other would step through registers with the times. The times would be sequentially fed into the step timer's time interval register. A system for variable interval times is shown in figure 20–13. The system shown is one that could apply to the dishwasher problem of figure 20–8.

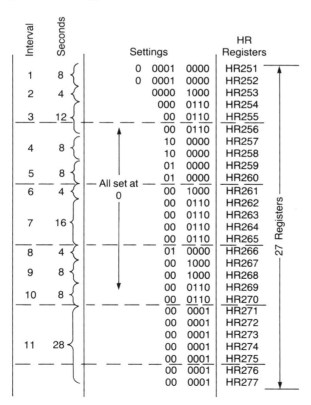

Figure 20-11
Sequencer
Multiple Time
Intervals

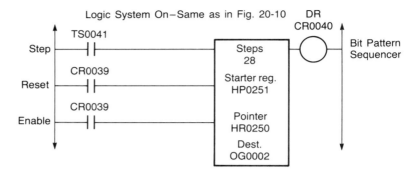

Figure 20-12
PC Program for
Figure 20-11

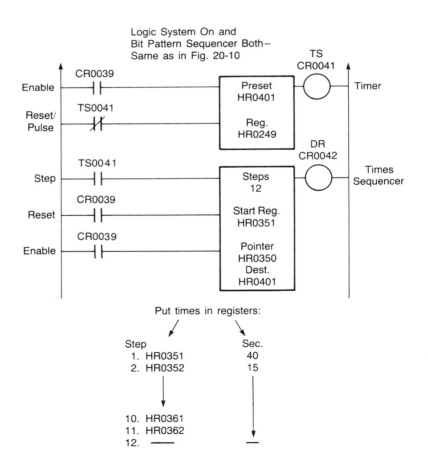

Figure 20-13
Sequencer
Interval Timer
Scheme

FORMAT WITH TIMES INCLUDED

Some manufacturers' sequencer formats have four options available, as shown in figure 20–14: time-driven SQO, event-driven SQO, time-driven SQI and event-driven SQI. SQO is a sequencer system that turns outputs on or off, depending on whether a 1 or 0 is put out to a given output.

Time-Driven SQO

Step	External Outputs (Data Entry: On=1, Off =0)					Dwell Time (PR Value)
	A	B	C	D	E	
1	Off	Off	Off	Off	Off	5 seconds
2	Off	On	Off	On	Off	20 seconds
3	On	On	On	On	Off	60 seconds
4	On	Off	Off	On	On	10 seconds

Beginning with the sequencer reset: When the SQO instruction goes True, step 1 is initiated; outputs A through E are Off. After a dwell time of 5 seconds (assuming SQO remains True), step 2 begins; outputs B and D go On. After 20 seconds, step 3 begins; outputs A and C go On. After 60 seconds, step 4 begins; outputs B and C go Off, output E goes On. After 10 seconds, a completion bit is set On and the cycle repeats with step 1.

Time-Driven SQI

Step	External Inputs (Data Entry: On = 1, Off = 0)					Dwell Time (PR Value)	Input-satisfied status bit
	A	B	C	D	E		
1	Off	Off	Off	Off	Off	120 seconds	Each step: this
2	Off	On	Off	On	Off	60 seconds	bit is ON only
3	On	On	On	On	Off	60 seconds	if inputs match programmed
4	On	Off	Off	On	On	120 seconds	data.

This sequencer moves from step to step in the same way as the time-driven SQO sequencer.

During the time that a particular step of the SQI instruction is in effect, the input-satisfied status bit will be set On only when the status of external inputs A through E matches the programmed input data for that step.

Event-Driven SQO

Step	External Outputs (Data Entry: On=1, Off=0)					False-True Transitions (PR Value)
	A	B	C	D	E	
1	Off	Off	Off	Off	Off	1
2	Off	On	Off	On	Off	1
3	On	On	On	On	Off	1
4	On	Off	Off	On	On	1

The PR value is set at 1 for each step (the typical case). Beginning with step 1, outputs A through E are Off. After a False-True transition of the SQO instruction occurs, step 2 is in effect; outputs B and D go On. After a 2nd transition, step 3 is in effect; outputs A and C go On. After a 3rd transition, step 4 is in effect; outputs B and C go Off, output E goes On. After a 4th transition, a completion bit is set On and the cycle repeats with step 1.

Event-Driven SQI

Step	External Inputs (Data Entry: On=1, Off=0)					False-True Transitions (PR Value)	Input-satisfied status bit
	A	B	C	D	E		
1	Off	Off	Off	Off	Off	1	Each step: this
2	Off	On	Off	On	Off	1	bit is ON only
3	On	On	On	On	Off	1	if inputs match programmed
4	On	Off	Off	On	On	1	data.

This sequencer moves from step to step in the same way as the time-driven SQO sequencer.

During the time that a particular step of the SQI instruction is in effect, the input-satisfied status bit will be set On only when the status of external inputs A through E matches the programmed input data for that step.

Figure 20–14
Four Sequencer Options

SQI is a comparison function. If the output, 1 or 0, matches the actual output status 1 or 0, a sequencer bit is set—1 for match, 0 for mismatch. The first two options in figure 20–14 operate an output group in the same manner as the format illustrated in figure 20–10. The time-driven SQO function shown can have the time interval programmed into the same register (address) as the output pattern. There is no need to have two sequencer functions, only one for both pattern and time.

The event-driven SQO function is essentially the same as the sequencer shown in figure 20–9: an input ladder-line change from off to on causes the function to increment to the next step and register.

The time-driven SQI and the event-driven SQI combine SQO sequencer functions and COMPARE functions in turning one logic bit off and on. They both examine the status of all active inputs. The input bits are compared to a set group of status bits: if the input bits exactly match the master set pattern, an input-satisfied bit is turned on; when one or more bits do not match, the input-satisfied bit will be off. The third SQI option function is time-driven, as is option one, and the fourth SQI option is event-driven, as is option two.

The program function that controls these sequencers is shown in figure 20–15 in ladder form rather than block form, as was the first format. Rungs 1, 2, and 3 control the steps. Rungs 3 and 4 tell the function where to stop. Rung 6 is used to reset the function back to the beginning at the first register.

Figure 20-15
Sequencer
Ladder Control
Diagram

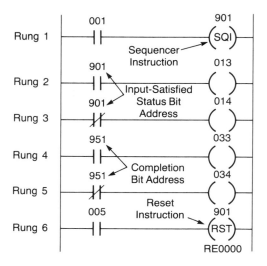

Ladder Diagram

CHAINING SEQUENCERS

Basic sequence functions have size limits. The single-function output capability may be 8- or 16-bits wide, and 8 or 16 output terminals can control only up to 8 or 16 machines. If there are 29 machines to turn on and off, even a single 16-bit program will not work. To add more output capability, you must chain, or expand, the number of bits across. Some formats have a place in the program or an address register to expand the bit capability. When a code key or number is inserted, it will automatically expand the number of outputs. Other formats, like the one shown in figure 20–16, must have parallel functions programmed in. Figure 20–16 shows how 29 outputs can be controlled from a 16-bit function format. Two functions are run in parallel. The extra 13 bits are controlled by the second functional block.

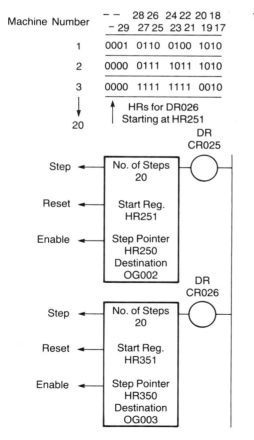

Figure 20–16 Expanded Sequencer Control for Many Outputs

Machine Number

	– – 28 26 24 22 20 18	16 14 12 10 8 6 4 2
	– 29 27 25 23 21 19 17	15 13 11 9 7 5 3 1
1	0001 0110 0100 1010	1111 1000 0101 1100
2	0000 0111 1011 1010	1111 0101 1001 0101
3	0000 1111 1111 0010	0000 0100 1010 0000

↓
20

↑ HRs for DR026
  Starting at HR251

↑ HRs for DR025
  Starting at HR351

DR
CR025

Step ← No. of Steps 20
Reset ← Start Reg. HR251
Enable ← Step Pointer HR250 Destination OG002

Both DRs are stepped from the same timing system. Reset and Enable circuits are also identical.

DR
CR026

Step ← No. of Steps 20
Reset ← Start Reg. HR351
Enable ← Step Pointer HR350 Destination OG003

The other sequencer dimension may need more steps for a process than one program block contains. Sequencers can have 64, 128, 256, and other numbers of steps per program block, depending on the format and PC model. Suppose there is a 128-step function limit for one basic function. If you need 277 steps, you must chain 3 functions together to run the process: 128 plus 128 plus 21 steps of the third function. Chaining is accomplished by starting the next sequencer after the last step of the previous one. A way to do this for one format is shown in figure 20–17.

Figure 20–17
Expanded Sequencer Control for Many Steps

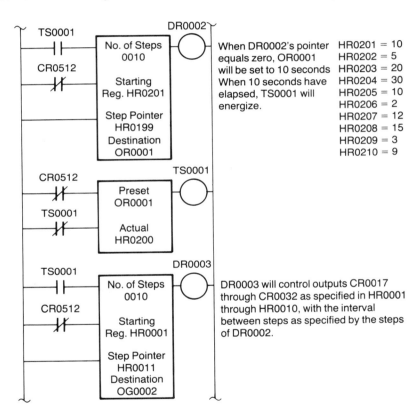

When DR0002's pointer equals zero, OR0001 will be set to 10 seconds. When 10 seconds have elapsed, TS0001 will energize.

HR0201 = 10
HR0202 = 5
HR0203 = 20
HR0204 = 30
HR0205 = 10
HR0206 = 2
HR0207 = 12
HR0208 = 15
HR0209 = 3
HR0210 = 9

DR0003 will control outputs CR0017 through CR0032 as specified in HR0001 through HR0010, with the interval between steps as specified by the steps of DR0002.

EXERCISES

1. Place the machine matrix shown in figure 20–18 in a PC Sequencer program. Program for manual, event-actuated operation.

2. Add individual times to the machine problem in exercise 1 for each step, using the same time interval for each step. Then, reprogram the problem using varying times, multiples, or variables, depending on your particular PC program format.

Step Number

Machine Number

Figure 20-18
Diagram for
Exercise 1

Step Number	7	6	5	4	3	2	1
1	—	On	—	—	—	—	—
2	—	—	On	On	—	—	—
3	On	—	—	—	On	On	On
4	On	On	On	On	On	On	On
5	—	—	—	—	—	—	—
6	On	—	On	—	On	—	On
7	—	On	—	On	—	On	—
8	—	—	On	—	—	—	On

Key — = Off

3. Create an operations scheme for a washer/dryer combination, using figure 20–8 for reference. Program a sequencer to run the wash/dry process. Make sure the program stops at the end of the cycle. It should not repeat until a reset switch is actuated. Note: do not use actual times in minutes for the program. Use seconds. Otherwise, program check-out takes a half hour or longer.

4. Chain a PC for chained outputs for 4 or 5 intervals. Exceed the number of output bits for one program block. Example, if the limit is 8, use 11. Choose your own pattern arrangement.

5. Chain a PC for chained steps for 4 or 5 output bits. Exceed the number of output steps for one program block. Example, if the limit is 128, use 135. Again, choose your own bit patterns.

6. Extra Credit: combine the system of exercises 4 and 5 and chain both output bits across and both output steps down.

Matrix Functions

21

At the end of this chapter, you will be able to

☐ Define AND, OR, and EXCLUSIVE OR gates (review from chapter 10).
☐ Define the additional concepts of COMPLEMENT and COMPARE.
☐ Describe the PC matrix construction system in register form.
☐ Describe and program the following matrix functions:
  AND, OR, XOR, COMPLEMENT, COMPARE.
☐ Use matrix functions in combination to simulate combination gates such as NAND and NOR.

INTRODUCTION

The word *matrix* can bring to mind a complicated and tedious mathematical procedure involving determinates, comparisons, cross multiplication, and other time-consuming operations. The PC matrix system eliminates the complication by enabling you to do a large number of comparisons or logic operations in a concise and orderly manner. The numbers involved in regular matrix algebra can be any decimal value: 13, −28, 45.782 or 134567.2, and so on; the PC matrix system involves only 1's and 0's. Furthermore, the PC matrix system does not involve cross multiplication. It is a special method for handling bulk data manipulations.

The PC matrix works with bits in one or two matrices and produces one resulting matrix. Chapter 10 covered digital gates. This chapter applies the various digital gates in large groups.

THE AND MATRIX FUNCTION

Four coils, each of which can be energized by two inputs in series, would result in four AND situations. The coils would be programmed on the PC in the usual manner shown in figure 21–1.

The upper section of figure 21–2 shows how the original four coils and eight inputs would be arranged in a 2-by-2 matrix. A 1 will represent on and a 0, off, in the conventional manner. Each bit of matrix A is used

Figure 21-1
Four Outputs
with Two Series
Inputs Each

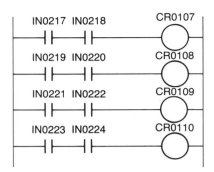

with the corresponding bit of matrix B. The bits, when used in a PC AND matrix, are analyzed for an AND situation. The result of the analysis is put in the corresponding bit location of matrix C.

In actual operation, the input data, or status, is contained in two series of registers. The AND operation for the two series then takes place and the results are put in another series of registers. This data is then moved to output registers. The equivalent register operation is shown in the lower section of figure 21-2.

Figure 21-3, a matrix AND operation, assumes some of the inputs on and some off for figure 21-2. The resulting outputs shown are determined by the PC multiple AND analysis.

You have used a 2-by-2 matrix for four AND functions. Next, suppose you had 53 coils, each with two series inputs for actuation. It would take a long time and a lot of PC memory to program the 106 contacts and the 53 coils. Using the AND matrix system makes programming a lot more straightforward. A typical PC uses sixteen 16-bit registers to give a 256-bit

Figure 21-2
Matrix Arrange-
ment for Figure
21-1

Figure 21-3
Two-by-Two
AND Analysis
for Figure 21-2

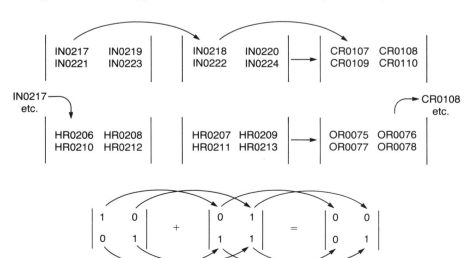

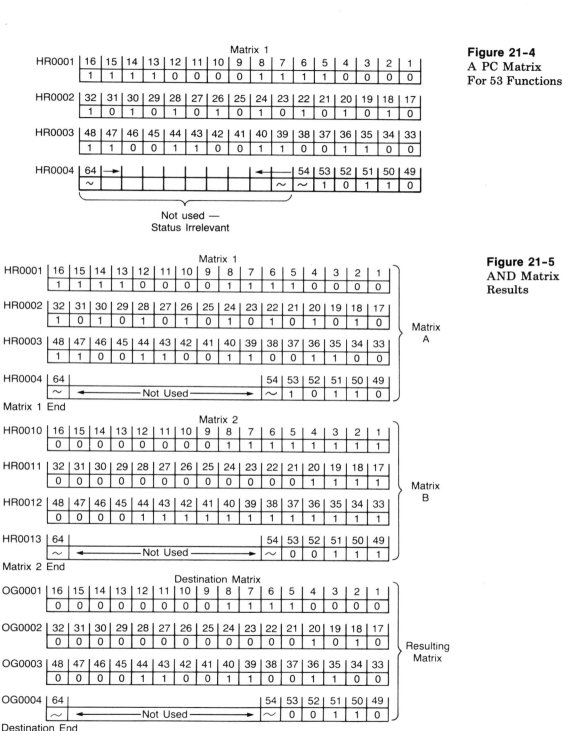

Figure 21-4
A PC Matrix
For 53 Functions

Figure 21-5
AND Matrix
Results

matrix. Using two input register matrices and one output matrix does up to 256 AND functions at once. This illustration uses only 53 of the 256 available register bits and four registers.

Figure 21-4 shows how to use 4 of the 16 registers available to perform the 53 AND functions. Tell the PC how many registers will be used in the operation; in this case four. We will use three full registers for the first 3 × 16, or 48, bits. The last 5 bits go in the first part of the next (fourth) register. This leaves 11 unused bits in the fourth register. This way, the other 13 registers are not involved, thus saving memory.

Assume some on and some off status for the 53 input AND matrices. The results of the matrix operation will appear in another matrix, as shown in figure 21-5.

How is the PC programmed to do the PC AND operation? Figure 21-6 shows a typical PC AND function. The coil is assigned a number in the usual manner. The AND function is carried out when the input is turned to the On status, as in most other functions. The illustration shows a general block configuration and a typical, specific, programmed AND function. In figure 21-6, you must also tell the PC which registers to use by

Figure 21-6
Typical PC
AND Matrix
Function

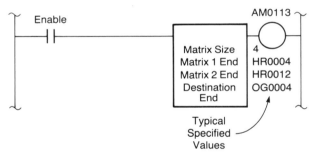

Figure 21-7
Typical
Allowable PC
Matrix Size by
Register Type

Type	Limit
HR	≤ 1792
IR	≤ 32 (PC-700) ≤ 8 (PC-900A) ≤ 16 (PC-900B)
OR	≤ 32 (PC-700) ≤ 8 (PC-900A) ≤ 16 (PC-900B)
IG	≤ 16 (PC-700) ≤ 8 (PC-900A/B)
OG	≤ 32 (PC-700) ≤ 8 (PC-900A) ≤ 16 (PC-900B)

Various PC Model Numbers

specifying the last register of each group of inputs and the output. In this case, the registers used for each part will be the one specified plus the previous three.

The coil of all matrix functions goes on when the functional operation is completed.

Matrix size varies among manufacturers. Maximum allowable size also varies, depending upon the type of registers used. Figure 21–7 shows one manufacturer's permissible variations.

THE OR MATRIX FUNCTION

The OR matrix operates similarly to the AND matrix, except that the bits in two matrices are compared on an OR logic basis instead of by an AND analysis. Figure 21–8 uses the same 53 on and off patterns that were

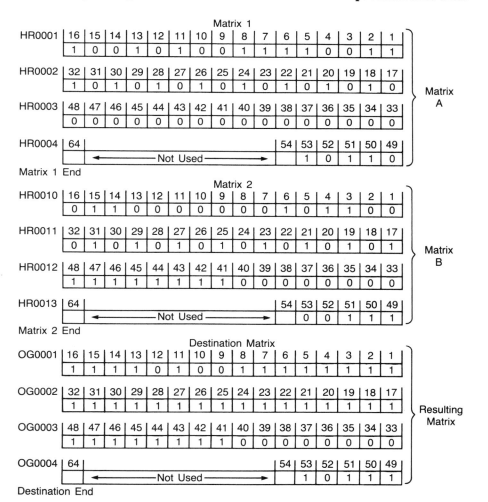

Figure 21-8
OR Matrix
Results

Figure 21-9
OR Matrix
Function

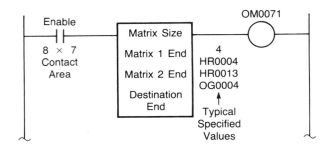

Figure 21-10
XOR Matrix
Results

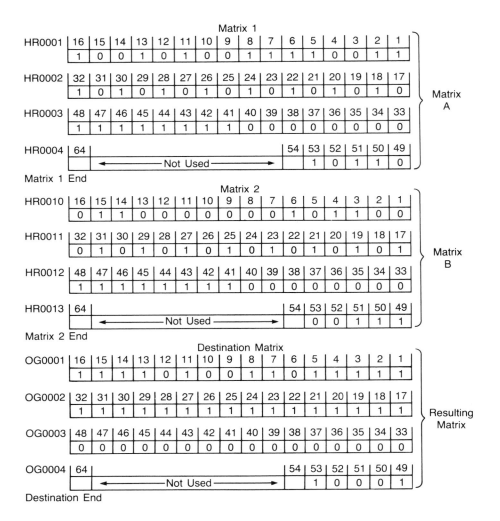

used for A and B in the AND example in figure 21–5. The results in matrix C are now determined on an OR basis instead of AND.

Figure 21–9 is a typical OR function layout. The size limitation of the matrix would vary in the same manner as the AND, as shown in figure 21–7. Operational procedure is essentially the same.

THE EXCLUSIVE OR FUNCTION

The EXCLUSIVE OR gate (XOR) is somewhat like the OR function, except that the output is not on when both inputs are on. The output is off when neither input is on, as in the OR function, and the output is on when either input is on, as with the OR function. However, when both inputs are on, the output is off. Figure 21–10 uses the same 53 inputs as in previous examples. The output matrix in this XOR example shows how the XOR is applied

The PC function for XOR is shown in figure 21–11. Size limits and operation are the same as for the AND and OR functions.

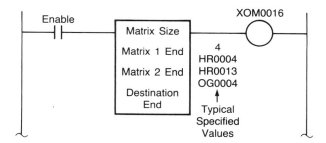

Figure 21–11
XOR Matrix
Function

THE COMPLEMENT FUNCTION

In some cases in a long PC program, you may wish to turn a number of devices to their opposite state. The COMPLEMENT function allows you to do so. All devices that are on can be turned off, and vice versa. Effectively, it changes all 1's in a matrix of applicable registers to 0 and all 0's to 1. Figure 21–12 shows the result of complementing register A, from our previous example, to register C.

The PC function for COMPLEMENT is shown in figure 21–13. Limits and operation are again the same.

THE COMPARE MATRIX

Another form of matrix logic is the COMPARE function, which compares two bits. If they are the same, it outputs a 1. If the original bits are different, it outputs a 0. Figure 21–14 is a truth table for this function.

Figure 21-12
COMPLEMENT
Matrix Results.

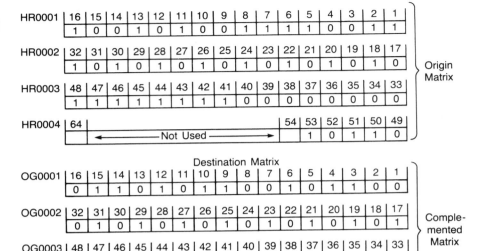

Figure 21-13
The COMPLE-
MENT Function

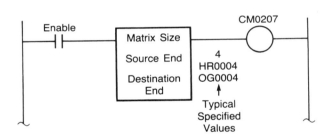

Figure 21-14
COMPARE
Function Truth
Table

Inputs		Output	Same?
A	B	C	
0	0	1	Yes
0	1	0	No
1	0	0	No
1	1	1	Yes

Again using the 53 bits in the matrix from the previous example, compare matrix A bits with those in B, as shown in figure 21-15. Matrix C shows which are the same and which are different by putting out 1 or 0, respectively.

The PC function for COMPARE is shown in figure 21-16. The usual size and operation description again applies. A frequently used form of

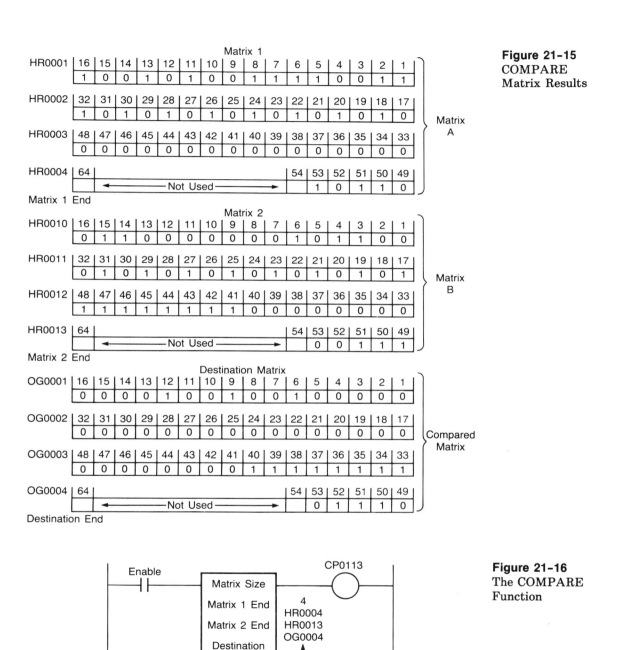

Figure 21-15
COMPARE
Matrix Results

Figure 21-16
The COMPARE
Function

the COMPARE function is the SEARCH matrix, which operates similarly to COMPARE. Your PC might have a different name for this type of operation; check your operations manual.

COMBINATION MATRIX OPERATIONS

Several matrix operations can be combined to perform special functions.

The NAND gate is an AND gate with an inverted output. To perform a matrix NAND, program two matrix operations in series, as shown in figure 21-17. First the AND matrix operation is performed from two input matrices. The result is placed in an output matrix. Next, the AND output matrix is complemented by the COMPLEMENT matrix function. The final output from the COMPLEMENT function is the NAND result for the original two matrices.

The NOR matrix operation is also a combination of two consecutive matrix operations. The results of an OR matrix are run through a COMPLEMENT matrix. The complemented resulting values are the NOR result of the original two matrices. Figure 21-18 shows how this function is programmed. The process is similar to the previously described NAND function.

Figure 21-17
The NAND
Function
Combination

Figure 21-18
The NOR
Function
Combination

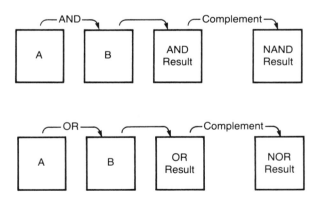

Inversions of inputs are accomplished by using a COMPLEMENT function on the front end of the matrix operation. If only a portion of the inputs must be inverted before use, other detailed operations are needed. These might be a number of MOVE functions or other data move systems available in the PC function list. The EXCLUSIVE OR function may be used to perform a selective complementation.

EXERCISES

The following two matrices are to be used for exercises 1 through 7:

Matrix A								Matrix B							
1	0	0	1	1	1	1	0	0	1	1	0	1	0	0	1
0	0	1	1	0	0	0	1	1	1	1	1	1	1	1	1
1	1	1	0	0	1	1	1	1	0	1	0	1	0	1	1
0	1	0	1	0	0	0	1	0	0	1	1	1	1	0	0
1	1	0	1	1	1	0	0	1	0	1	0	1	0	1	1
0	0	0	1	0	0	1	0	1	1	0	1	0	0	1	0
0	0	0	0	0	0	0	1	1	0	1	1	0	0	0	1
1	0	1	0	1	0	1	0	0	1	0	0	0	1	1	1

1. AND A and B to determine matrix C.
2. OR A and B to determine matrix D.
3. XOR A and B to determine matrix E.
4. COMPARE A and B to determine matrix F.
5. COMPLEMENT matrices A, C, and E to determine matrices G, H, and J.
6. NAND B and C to determine matrix K.
7. NOR A and C to determine matrix L.

Examine the following examples by programming them on a PC:

8. Figure 21–5, AND.
9. Figure 21–8, OR.
10. Figure 21–10, XOR.
11. Figure 21–12, COMPLEMENT.
12. Figure 21–15, COMPARE.
13. Figure 21–17, NAND.
14. Figure 21–18, NOR.
15. Create two matrices of your own design with 43 active bits. In a 16-bit PC, matrix analysis will require two full registers plus part of a third. Find the solutions to functions of your choice manually and by the PC. Compare the results for exact correspondence.

SECTION SIX

ADVANCED FUNCTIONS

Controlling a Robot With a Programmable Controller

22

At the end of this chapter, you will be able to

☐ Develop a "coil and contact" (input/output) control system to operate a basic robot.
☐ Develop a drum controller/sequencer control system to operate a basic robot.
☐ Describe the electrical connection and interfacing system required to connect a PC to an industrial robot.
☐ Develop a drum controller/sequencer control system to operate an industrial robot.

INTRODUCTION

This chapter will illustrate how a PC may be used to run a robot. The robots to be used for illustration are *pick-and-place* robots which have various discrete positions for their gripper assembly. The positions are determined by discrete signals—"ON" causes the robot's axes and manipulators to move to one extreme position, and "OFF" moves them to the other extreme position. The other type of robot, the *continuous path* type, is not discussed in this chapter. This type must be controlled by analog computers, or PID (see chapter 24) control devices, which are much more complicated and are not discussed in detail in this text.

The chapter progresses through various levels of complexity of pick-and-place robot control. First, on-off switches will be used to manipulate a basic robot. Second, a drum controller/sequencer will be used for control of the same robot. Then, PC control systems will be used for industrial-type pick-and-place robots.

A BASIC TWO-AXIS ROBOT WITH GRIPPER

We will progress through two schemes of controlling the basic pick-and-place robot shown in figure 22-1. We will use switches first, and then a drum controller/sequencer.

Figure 22-1
Basic Pick-and-
Place Robot

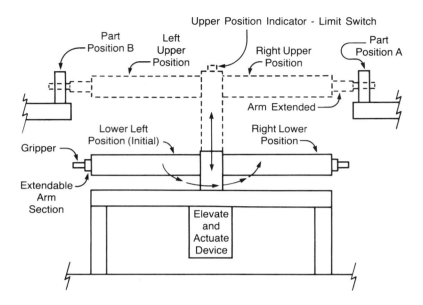

The robot used for illustration starts operating from the position shown, which is the "at-rest," lower-left, initialized position. The step-by-step sequence of operation to move a part from position A to position B is as follows:

1. Arm is initially in the down-left position as shown. Gripper is open and not extended.
2. Arm moves to upper position
3. Arm rotates to right
4. Hand extends to position A
5. Gripper closes, gripping part
6. Arm swings back to the left to position B
7. Gripper opens, releasing part
8. Hand retracts
9. Arm lowers to the initial position

For illustration, assume that the robot has four powered pneumatic solenoids. If all solenoids are off, no air is applied to the robot's actuators. In this initial position, the robot is in the lowered, left position with the hand retracted and the gripper open. Energizing each of the four solenoids causes the following action to occur:

1. ROTATE—Arm rotates full right
2. RAISE—Arm rises to the upper position

3. EXTEND—Hand extends from the arm

4. GRIP—The gripper closes

More than one solenoid can be energized in combination to facilitate operation. If a solenoid is not energized, the function is in the other extreme initial position, opposite those listed.

An operational matrix for the robot to move a part from position A to position B is shown in figure 22–2.

Step	Up	Rotate Right	Hand Out	Grip Close
Initialized	O	O	O	O
1	X	O	O	O
2	X	X	O	O
3	X	X	X	O
4	X	X	X	X
5	X	O	X	X
6	X	O	X	O
7	X	O	O	O
8	O	O	O	O

Figure 22-2
Part Movement
Robot
Operational
Matrix

An "O" indicates the opposite position; down, left, in, or open.

PC SEQUENCE CONTROL OF A BASIC PICK-AND-PLACE ROBOT

A simple control system for the robot shown in figure 22–1 could consist of four switches, one for each motion. One disadvantage of the four-switch control is that someone would have to do the controlling continuously. In addition, turning off a switch would not immediately stop the arm; it would spring-return to its initial position, which would be hazardous to anyone expecting it to stop immediately.

There could also be problems in mechanical interferences during operation. In the upper position, with the arm extended, moving the arm down could break off the arm on the conveyor below it. Also, if the gripper opened up while the arm was making a swing, the part would be dropped or thrown outward.

For these and other reasons a ladder diagram with interlocks and sensors included for the robot's control should be developed. We will not develop the complete ladder control system in this chapter, but we will develop a drum controller/sequencer program of the type commonly used for robotic control.

Two basic programs for controlling the robot of figure 22–1 is shown in figure 22–3. The first program is a PC version of the switch/relay system. The second is a DR function and registers. The DR is step-pulsed at intervals by a timer. The timer's preset times would be determined by the interval or intervals required for each operation to be completed. As an example, if the arm swing time takes four seconds, the time interval before starting the next step should be five seconds or more.

Again for this program, there are some possible operational problems. If a pneumatic cylinder failed or something became jammed, the PC program would continue unabated. There could be equipment damage or even

Figure 22-3
PC Program for
Robot Control

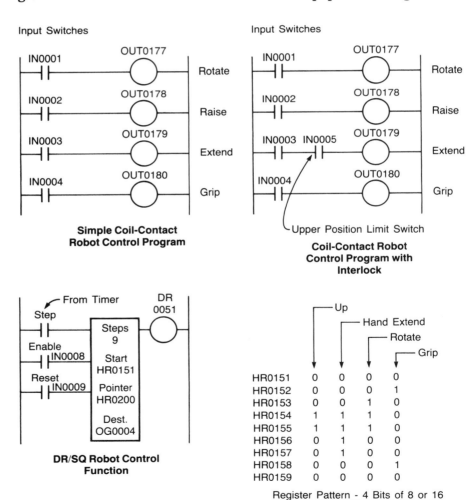

personnel injury. Additional programming would be necessary to include interlocks, sensors, positive emergency stops, and the like. The procedure for developing additional programming was covered in chapters 6 and 11.

A MORE COMPLICATED ROBOT

The robot in the previous section would more accurately be called *fixed automation*. Several different MOVE sequences would have made it a true robot. If, for example, you occasionally moved the part from B to A, instead of A to B, you would need two programs with different on-off patterns. The ability to change quickly from one program to another makes a system robotic. The insertion of different register patterns could be accomplished by MOVE functions. Note that a robot is loosely defined as a *reprogrammable manipulator*.

Next, consider the robotic control for the work cell shown in figure 22–4. The parts may come in on conveyor A or conveyor B and go out on either A or B. There are three possible process operations in the work area: drill 1/2 or 3/4 inch, counter-sink, and counter-bore. A drilling operation for one of the two sizes is always done; counter-sink and counter-bore may or may not be included for any particular part.

For this process, then, there are 32 possible combinations of individual moves for a given part (5 alternates of 2 possible operations is 2 raised to the 5th power, or 32). This requires a program catalog of up to 32 different programs. The various possible combinations are shown in figure 22–5.

We will not write detailed PC programs for the work-cell operation. We will, however, show the PC programming system in block form, as shown in figure 22–6.

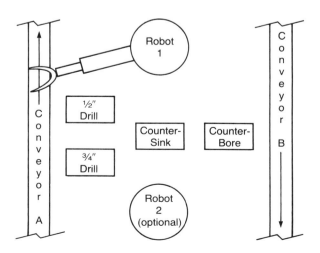

Figure 22-4
Work Cell with Drilling/Boring Operations

Figure 22–5
Combinations of
Operations for
Figure 22–4

In on Conveyor	Drill Size	Counter-Sink	Counter-Drill	Out on Conveyor	
A	¼″	Yes	Yes	A	
or	or	or	or	or	
B	½″	No	No	B	
	× 2 =	× 2 =	× 2 =	× 2 =	} Combinational
2	4	8	16	32	} Possibilities

Figure 22–6
Programming
Scheme for the
Work Cell

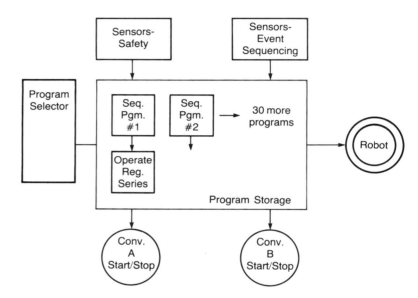

CONTROLLING AN INDUSTRIAL PICK-AND-PLACE ROBOT

An industrial-type robot is shown in figure 22–7. It has various motion
and gripping capabilities similar to those in figure 22–1. These are:

1. Arm up or down, elevate
2. Arm rotate 180 degrees
3. Gripper rotate 180 degrees
4. Gripper open and close
5. Gripper extend and retract
6. Slide left or right, 5 stations—2 ends, 3 intermediate

The robot shown in figure 22–7 is operated by applying 110 volts AC to
various pneumatic solenoids. Each solenoid, when energized, lets its air
valve supply air for operating the various functional motions. When off,
no air is supplied, and the function is in its initial position. As in the

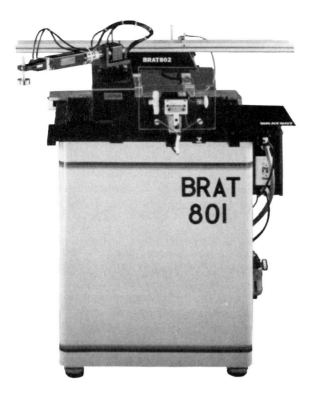

Figure 22-7
Industrial Pick-
and-Place Robot
(Courtesy of TII
Robotics)

previously discussed robots, the positions used are the extreme ones at one end of travel or the other. The only exception to end positions is motion 6: for slide right or left, there are three intermediate stopping points in addition to the two at the ends of travel.

This industrial robot differs from the previous one in that motions cease when a function is deenergized. It also differs in that some opposite motions require two inputs, one for each direction.

There are various control methods used to operate the robot or to program the robot to operate automatically. These include manual control, mechanical drum control, EPROM integrated circuit chip control, and computer control. Computer control may be further subdivided. One computer type has a teach pendant, which is used in conjunction with a computer memory. It can record, remember, and repeat a series of steps performed manually. The other type is a straight computer program.

Also, many robotic devices can be controlled from a small computer (a master program is entered from a disk or computer memory). However, an increasing number of robots, like the one in figure 22-7, are being controlled from programmable controllers.

The illustrated robot could be supplied with a PC control package built in by the robot manufacturer. The next section illustrates how to use your own PC to interface with the robot shown.

CREATING A PC ROBOT CONTROL SYSTEM

Before programming the PC to control the robot, you must develop a scheme to connect and interface with PC with the robot. Figure 22–8 shows the pin connections to the robot and the necessary color code/wire numbers of the connecting cable. Since the robot uses 110 volts, you need a 110 volt interface I/O for PC inputs and outputs. If you connect the ground, and group common connections by direct wiring, the PC needs only 13 output ports. You therefore would choose a 16-output PC output module.

Figure 22–8
Robot Control
Cable Pin

Below is a listing of the I/O numbers and the letter that corresponds to the I/O on the cable between the I/O rack and control panel designed for the PC. Also listed are the robot pin # and the corresponding function.

PC Input #	Cable Letter	Robot Pin #	Robot Function
1	A	—	—
2	B	—	—
3	C	—	—
4	D	—	—
5	E	—	—
6	F	—	—
7	G	—	—
8	H	22	Aux. Input
9	J	23	Aux. Input
10	K	24	Aux. Input
11	L	25	Aux. Input
12	M	2	Station 1
13	N	3	Station 2
14	P	4	Station 3
15	Q	5	Station 4
16	R	6	Station 5
	S	7	Common Input

PC Output #	Cable Letter	Robot Pin #	Robot Function
17	T	8	Grip
18	U	10	Elevate
19	V	15	Extend
20	W	11	Rotate CW
21	X	12	Rotate CCW
22	Y	13	Slide Right
23	Z	14	Slide Left
24	a	9	Rotate Grip
25	b	19	Aux. Output
26	c	20	Aux. Output
27	d	21	Aux. Output
28	e	—	—
29	f	—	—
30	g	—	—
31	h	—	—
32	j	—	—
	k	16	Common Output

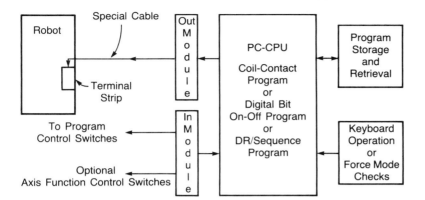

Figure 22-9
Robot–PC
Control System
Block Diagram

Figure 22-9 shows a block diagram of this control scheme. Note that interlocks will not be considered in this discussion, but would be included in advanced programs.

Connections/Interconnections

Now that the PC and the robot are connected, you can choose a programming format. You may choose a contact and coil control format, which was explained earlier in this chapter, or you could use a digital bit program similar to those in chapter 19. This illustration will use the drum/sequencer discussed in chapter 20, which is the most commonly used format.

A DRUM/SEQUENCER PROGRAM FOR THE ROBOT

Figure 22-10 is a program constructed on a program coding sheet. Blank coding sheets are furnished by the robot manufacturer for ease of program formulation. The sequential steps of operation go from top to bottom. A step description is written on the left of the sheet under *Sequence of Events*. The off-on pattern is indicated by X's for on and a blank space (or sometimes O's) for off. A listing of functional status for each motion is listed along the top (Rotate CCW, Rotate CW, etc.).

The sequence of events to accomplish the operational objectives was determined and listed. Then, the X's were filled in for the functions that will be actuated. Note that in many cases, more than one function must be on at a given step. The final part of the procedure was to record the assigned PC operation numbers for each step (28, 27, etc.). Our program uses the output group 2 register, OG2. Therefore, the individual 16 outputs are 17 through 32. A cross reference of cable numbers and letters corresponding to operational numbers was shown at the bottom of figure 22-8.

Figure 22-11 shows a typical PC program that could be used for the robot sequencer program. The programming follows the system of chapter

Figure 22-10 Program Code Sheet for Programmable Controller (Courtesy of TII Robotics)

Sequence of Events	Time Interval	Step No.	28 ROTATE CCW	27 ROTATE CW	26 UP	25	24 SLIDE RIGHT	23	22 GRIP	21 SLIDE LEFT	20 EXTEND	19	18	17 GRIP ROTATE
HOME POSITION (INTIALIZE)		H												
ROTATE CW SLIDE RIGHT		1	X				X							
SLIDE LEFT, MAN. UP		2			X					X				
MAN. UP, EXTEND		3			X						X			
MAN. UP, EXTEND, GRIP		4			X				X		X			
MAN. UP, EX., GRIP, ROT. CW		5		X	X				X		X			
GRIP, EX., GRIP ROTATE		6							X		X			X
EXTEND, GRIP ROTATE		7									X			X
GRIP ROTATE, MAN. UP		8			X									X
SLIDE RIGHT, MAN. UP		9			X		X							
EXTEND, MAN. UP		10			X						X			
GRIP, EXTEND, MAN. UP		11			X				X		X			
GRIP, EX., MAN. UP, ROTATE CCW		12	X		X				X		X			
GRIP, EXTEND		13							X		X			
EXTEND		14									X			
NEUTRAL POSITION		15												
SLIDE LEFT, GRIP ROTATE		16								X				X
EXTEND, GRIP ROTATE		17									X			X
GRIP, EXTEND, GRIP ROTATE		18							X		X			X
MAN. UP, SLIDE RIGHT, GRIP, EX.		19			X		X		X		X			
MAN. UP, EXTEND		20			X						X			
MAN. UP		21			X									
ROTATE CW, GRIP ROTATE		22		X										X
EXTEND, GRIP ROTATE		23									X			X
GRIP, EXTEND, GRIP ROTATE		24							X		X			X
MAN. UP, GRIP, SLIDE LEFT, EX		25			X				X	X	X			
		26												
		27												
		28												
		29												
		30												

Switch No. / Step No. / Time Interval / Sequence of Events; ← Station

20. It is operated on a 5-second time interval, as set in the timer program. Note that the pattern of regular bits corresponds to the coding sheet of figure 22-10. Only the first 12 bits are needed, since there are only 12 functions to be controlled.

The program shown in figure 22-10 is time based. In actual operation, for equipment and personnel safety, the program could be all or partially event based. There would be limit switches or sensors for position or part-in-place indication. An ensuing operational step would not be able to start until the previous step is successfully completed. For example, if there were no part to be picked up, the sequence (not shown) would stop or the arm would return to home position.

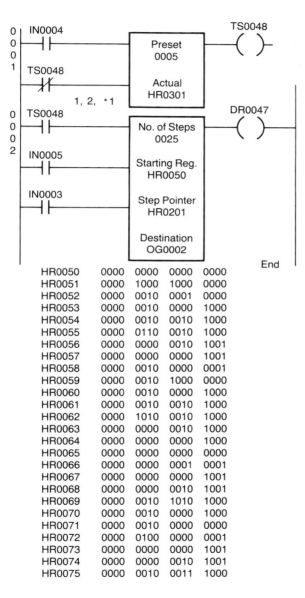

0	IN0004					TS0048

Figure 22-11
PC Robot
Control Program
and Register
Pattern

Preset
0005

Actual
HR0301

1, 2, *1

TS0048

No. of Steps
0025

DR0047

IN0005

Starting Reg.
HR0050

IN0003

Step Pointer
HR0201

Destination
OG0002

End

HR0050	0000	0000	0000	0000
HR0051	0000	1000	1000	0000
HR0052	0000	0010	0001	0000
HR0053	0000	0010	0000	1000
HR0054	0000	0010	0010	1000
HR0055	0000	0110	0010	1000
HR0056	0000	0000	0010	1001
HR0057	0000	0000	0000	1001
HR0058	0000	0010	0000	0001
HR0059	0000	0010	1000	0000
HR0060	0000	0010	0000	1000
HR0061	0000	0010	0010	1000
HR0062	0000	1010	0010	1000
HR0063	0000	0000	0010	1000
HR0064	0000	0000	0000	1000
HR0065	0000	0000	0000	0000
HR0066	0000	0000	0001	0001
HR0067	0000	0000	0000	1001
HR0068	0000	0000	0010	1001
HR0069	0000	0010	1010	1000
HR0070	0000	0010	0000	1000
HR0071	0000	0010	0000	0000
HR0072	0000	0100	0000	0001
HR0073	0000	0000	0000	1001
HR0074	0000	0000	0010	1001
HR0075	0000	0010	0011	1000

TS0048

0
0
0
1

0
0
0
2

267

EXERCISES

1. Refer to the robot in figure 22-1. Develop a pattern similar to that in figure 22-2 for a different sequence of operation. You must move a part from position B to position A. The robot starts at the lower-left, initialized position.

2. For exercise 1, develop a drum sequencer program to move the part from B to A.

3. The work station/robot in figure 22-12 is similar to the one in figure 22-1. It differs in that it has 6 active positions, not 4. It has a positioning solenoid for each of the 6 positions shown. It also has arm-extend and gripper-close actuators, as in figure 22-1. Develop a coil-and-contact PC program to accomplish moving a part from LM to UR. Also develop a program to move a part from UR to LL.

Figure 22-12
Diagram for
Exercise 3

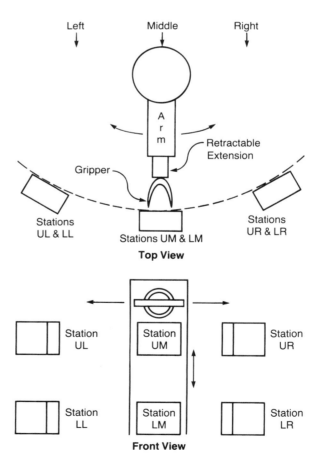

4. The robot shown in figure 22–13 is similar to the robot in figure 22–7. Develop a timed (times of your choice) sequence program to accomplish the following sequence below. An example program was illustrated in Figure 22–10.

   1. Initial position as shown

   2. Pick up part at A5

   3. Move part to B4

   4. Return to initial position

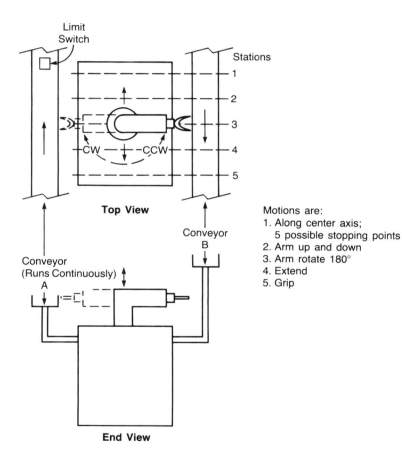

Figure 22-13
Diagram for
Exercise 4

Motions are:
1. Along center axis;
   5 possible stopping points
2. Arm up and down
3. Arm rotate 180°
4. Extend
5. Grip

5. Develop a timed sequence program to accomplish this longer program in the same manner as exercise 3.

   1. Initial position as shown

   2. Pick up part at B1

3. Move part to A4
4. Conveyor moves part along
5. Part reaches position A1
6. Limit switch actuated by part
7. Robot picks up part at A1
8. Part moved to B5
9. Return to initial position

For this exercise, two programs may be needed, separated by an LS1 actuation ladder logic line.

Analog PC Operation

23

INTRODUCTION

This text has so far dealt with discrete PC operation; input and output statuses have been on or off. This chapter will consider analog PC operation. Analog PC control can be used to control any process with variables as a control consideration. Many medium and large PCs are able to deal with analog signals as having discrete functions. For analog operation, the level of a PC input signal is sensed by an analog input module. In addition, the level of the output can be a variable value as sent to the process from an analog output module. The PC analog input capability enables you to monitor such devices as thermal indicators, pressure transducers, electrical potentiometers, and many other data input devices with varying signal values. Output PC analog control can be positioned at many intermediate positions. This control is in contrast to discrete control, which operates only at its two extremes.

BCD PC analog input and output value ranges are divided into a number of steps. BCD analog input devices include thumbwheels, encoders, and the like. Analog output devices control such devices as digital numbers, seven-segment displays, and stepper motors.

PC analog capabilities allow many different actions for one single input, depending on the input's value. For example, a process in which 20 lights are used to indicate how full a tank is in 5 percent increments would need only one analog input and one sensor; a discrete system would need

20 on/off sensors and 20 inputs. Analog output programs have similar advantages; for example, a single analog output can position a valve in many different positions.

Analog capability enables you to control continuous processes in such industries as chemical and petroleum. Any number of variable input signals can be received by a PC module and then processed mathematically by the CPU. The resulting analog value or values are then sent to an output module. The analog output module signal then controls a variable process or processes.

TYPES OF PC ANALOG MODULES AND SYSTEMS

Analog PC systems are of two general types: the BCD and the straight numerical.

The BCD analog PC system is sometimes called the multi-bit type. (Chapter 9 covered the BCD numbering system and its uses.) Figure 23-1 shows the operation of a thumbwheel input to an input BCD module. BCD codes are fed into the PC input module from the thumbwheel output. Other possible BCD-type inputs are bar-code readers, and encoders. A BCD output module is also shown in figure 23-1. In this case, BCD codes are fed from the output module to a numerical indicating device. BCD output devices include such things as digital number displays, variable position actuators, and stepper motors.

The other general PC analog system is the straight numerical type. Some typical ranges of the modules available for these systems are shown

Figure 23-1
Analog BCD
Input and
Output Systems

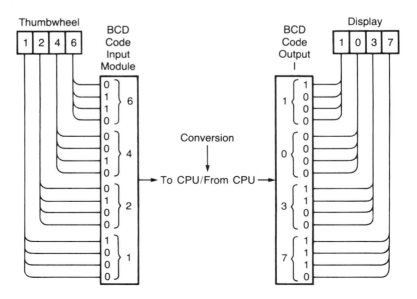

2-10 mA
4-20 mA
10-50 mA
0 to + 5 Volts DC
0 to + 10 Volts DC
± 2.5 Volts DC
± 5 Volts DC
± 10 Volts DC

Figure 23-2
Typical Analog
I/O Module
Ranges

in figure 23-2. The PC numerical type module is used for a large variety of input devices, the most common of which is the electrical potentiometer. The potentiometer is used to input a linear, varying, electrical value to the input module. The potentiometer can be one that reads either temperature, pressure, distance, position, or electrical values. Other inputs include thermocouplers, strain gauges, and straight electrical signals. The more-complicated analog system, the PID output system, will be briefly discussed in chapter 24.

Note that the PC handles continuous analog systems in discrete steps. The continuously varying input signal is not strictly continuous when it reaches the PC CPU. As we divide up the input signal into more steps, it more nearly approaches the exact duplication of the input signal. Figure 23-3 shows how the input signal is divided into more parts for increased accuracy. As the number of input divisions goes up from 8 to 16, the digital signal more nearly describes the actual input signal. Some normally used PC divisions are 1024 and 4000. As the number of divisions goes up, the PC system cost also goes up. You must have enough steps to control your process with precision, but not so many that cost becomes prohibitive.

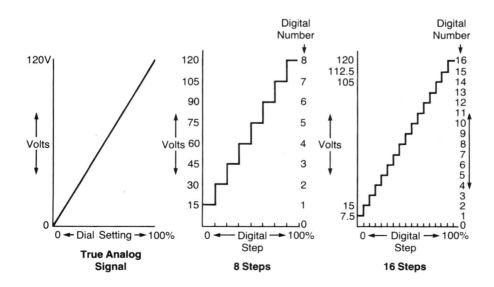

Figure 23-3
Analog Signal
Conversion to
Digital Steps

ANALOG SIGNAL PROCESSING

The sensor or signaling device that feeds the input module does not usually have the same electrical range as the input module. Its lower-limit electrical value must be matched to the lower-limit electrical value of the input module. The input's upper-limit signal value must also be matched to the upper-limit electrical value of the input module through the use of an intermediate signal conversion. Similarly, the output module and the outputs must have their signals appropriately matched by a converter. Intermediate values must also be linearly matched by the converters for both input and output.

The digital-coded signals sent from the analog input module to the CPU are proportional to the module's input electrical signal; therefore, the resulting digital code values are also proportional to the original input signals. Figure 23–4 shows a block diagram of the conversion process, along with two numerical examples. This module has 1024 steps, but it could have had 512, 4096, or some other power of 2. These are typical values for the module's divisions.

Figure 23–4
Analog Input
System

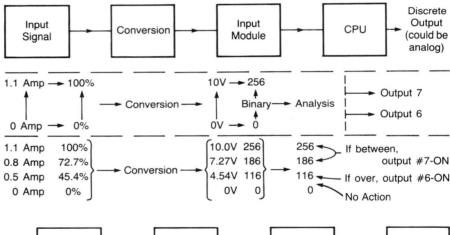

Figure 23–5
Analog Output
System

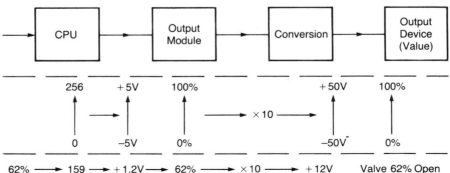

Suppose you need to compare the input values with a fixed value, or with another inputted value for action. Say, for example, that if the sensor signal exceeds 0.5 amps, you must turn on output 6. Further, if the sensor signal is between 0.8 and 1.1 amps, you must turn on output 7. In this case you have an analog input and a discrete output. The signal-level processing of these electrical values is shown in figure 23–4.

A more typical situation is analog in and analog out. Figure 23–5 shows an example of analog-to-analog process conversion. The figure assumes direct input-to-output conversion. As we shall subsequently see, the CPU can scale or modify the data as it passes through the conversion block.

BCD OR MULTIBIT DATA PROCESSING

BCD data is handled similarly to analog data. Figure 23–6 shows a block diagram of how BCD devices and data are used by the PC. The input and output devices are mathematically matched directly by the input and output modules. No conversion of values is required since the input and output devices are built to match the modules directly. In this illustration, the input number is entered directly from thumbwheels. The input data is scaled to half for illustration. The resulting half value is sent to the output device, a four-digit, seven-segment display. Since the CPU does math in binary, appropriate BCD and binary conversions are carried out as shown.

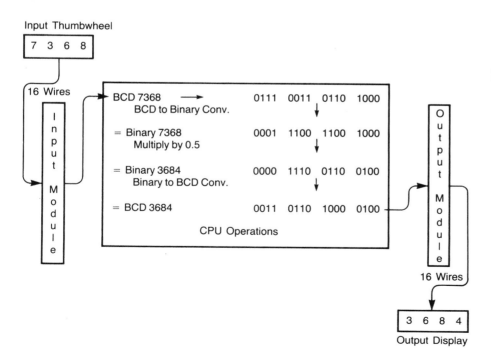

Input Thumbwheel

Figure 23-6
BCD (Multibit) Input and Output System

Output Display

CHAPTER EXAMPLES

The inputs and outputs of analog systems can be straight analog or BCD, as we have shown, or they can be other types such as thermocouple, load cells, humidity transducers, and electric motor drive input and feedback signals. In addition, discrete modules may be used in combination with analog modules. Furthermore, multiple inputs and outputs can be involved. To illustrate all of these combinations would require many examples.

This section will present six examples, four of which have multiple inputs or outputs, as representative of analog PC operations. These are:

☐ Example A: Analog in/discrete out
☐ Example B: BCD in/discrete out
☐ Example C: Analog in/analog or BCD out
☐ Example D: BCD in/BCD or analog out
☐ Example E: Two analog in/Two analog out
☐ Example F: Two BCD in/two analog out

Example A: Analog In and Discrete Out

The example given in figure 23–4 was analog in and discrete out. The figure's values further illustrate the example. The problem is to have an output go on when a certain level, 0.5 amps, is reached, and to have another output to be on when the amps are between the values 0.8 and 1.1 amps. To accomplish this, use the input values with comparison functions. The comparison functions were covered in chapter 16.

For the first output, use a GE function. Since the second condition has two limits, it needs two comparison functions; in this case, greater than and less than. Figure 23–7 illustrates how to program the PC to accomplish the required comparisons and energize appropriate outputs.

Example B: BCD In and Discrete Out

A problem similar to example A is illustrated in example B. The input is a BCD thumbwheel that counts up to 9999. If the input is 3750 or above, output 6 is to go on. If the input is between 6200 and 8542, output 7 is to go on. Assume that the input data is received in register IR 0006 in the CPU. As previously stated, the CPU works in binary; therefore, you must first convert the IR 0006 value to binary. Register HR 0045 will receive the converted value. Thereafter, the comparison functions are the same as in example A. Figure 23–8 shows the PC programming for this BCD comparison problem.

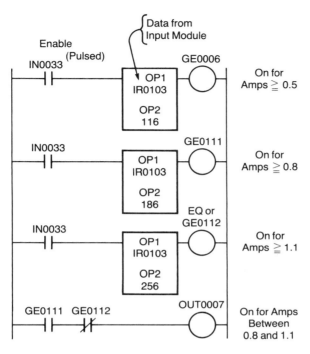

Figure 23-7
Example A:
Analog In/
Discrete Out

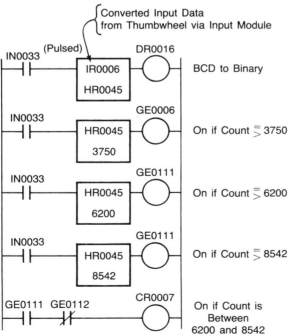

Figure 23-8
Example B:
BCD In/Discrete
Out

Example C: Analog In and Analog or BCD Out

For example C, an analog signal of 0 to 10 volts comes in through a converter to an input module. The signal is to be scaled to 1/5 of its value by the CPU, and then sent out through an output module. The output is also to be sent to a BCD output display. Figure 23–9 illustrates how the PC can be programmed internally to accomplish one or both output conditions. The analog signal goes out directly. The output signal to the analog output is first converted to BCD and then sent to the BCD display.

Figure 23–9
Example C:
Analog In/
Analog or BCD
Out A. Data
Flow Diagram/
B. PC Program

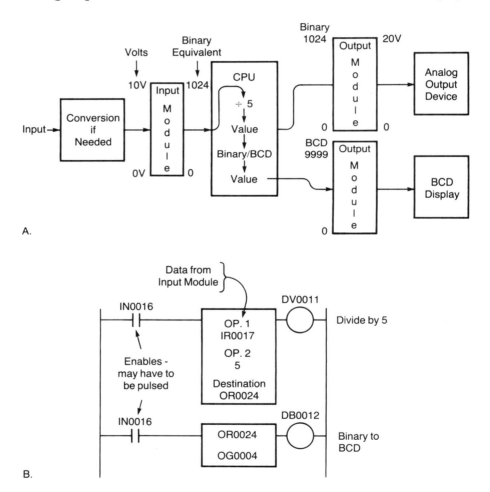

Example D: BCD In and BCD or Analog Out

In this example, a BCD input, 0 to 9999, is received by an input module, which places the value received into register IR 0004. A fixed value of 180 is to be subtracted from the value received, and the result is to be

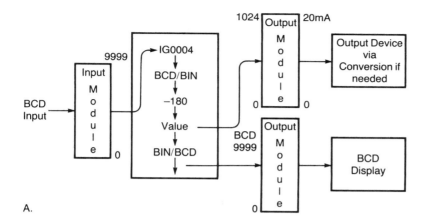

Figure 23-10
Example D: BCD In/BCD or Analog Out A. Data Flow Diagram. B. PC Program

A.

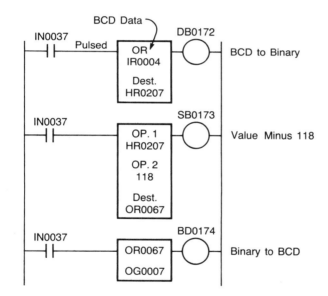

B.

sent out to a 0–9999 BCD output display. Additionally, the output value is to be placed in a 0 to 20 milliamp analog output module. Figure 23–10 illustrates the PC programming necessary to accomplish the transfer of the original input value, less 180. Appropriate BCD-to-binary conversions are included in the program.

Example E: Two Analog In, Mathematical Manipulation, Two Analog Out

Examples E and F both have multiple inputs and outputs. They could have had more than two inputs and a single output, but in this illustration both examples use two inputs and two outputs.

Figure 23–11
Example E: Two
Analog In, Add
and Subtract,
Two Analog Out
A. Data Flow
Diagram. B. PC
Program

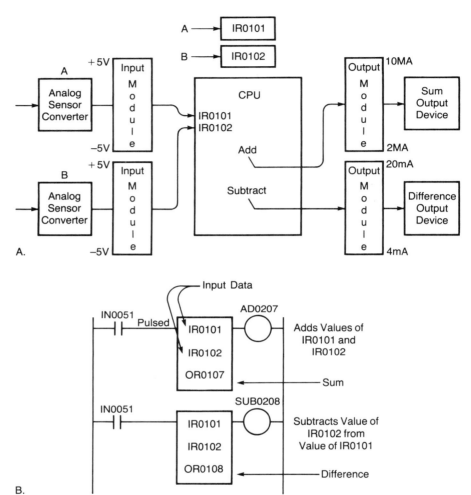

Example E in figure 23–11 has two analog inputs whose values are manipulated in the CPU. For illustration, the values are both added and subtracted. The sum is output to one analog output and the difference is sent out to another. The internal programming to perform the mathematical manipulations is also shown.

Example F: Two BCD Inputs, Mathematical Manipulation, Two Analog Outputs

Example F, shown in figure 23–12, is similar to example E. It has two BCD inputs and two numerical analog outputs. You could also mix and match analog and BCD inputs and outputs with little difficulty. In ex-

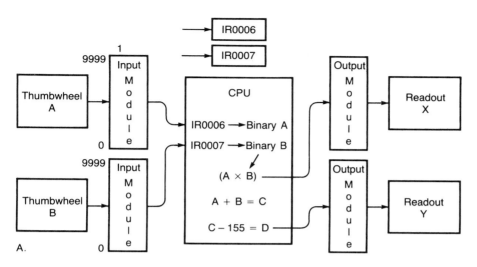

Figure 23-12
Example F: Two
BCD in, Multi-
plication, Addi-
tion, Subtrac-
tion, Two
Analog Out
A. Data Flow
Diagram. B. PC
Program

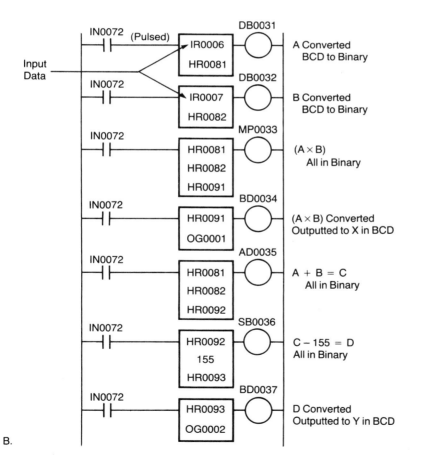

ample F, the first output is the product of the inputs and the second output is the sum, minus 155. The product is read on readout X and the sum, minus 155, on readout Y. Again, the PC program to perform the math functions is shown.

EXERCISES

1. Draw an output graph for a 32-step (5-bit) output similar to those shown in figure 23–3. 144 volts is 100 percent. The input configuration is the same as figure 23–3. Determine the output digital-step-indicated voltage (range) for dial settings of 23, 45, 46, and 78.5 percent.
2. If the input graph curve in figure 23–4 were nonlinear, as shown in figure 23–13, would the output be linear with respect to dial setting? Explain.

Figure 23–13
Diagram for Exercise 2

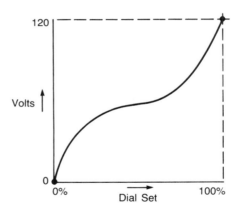

For exercises 3 through 5, assume:

- ☐ Input volts 0 to 80.
- ☐ Input module volts 0 to 5.
- ☐ Binary has 128 steps.

For exercises 3 through 5, include the following in each answer:

- ☐ Draw a block flow diagram as in figure 23–4.
- ☐ Draw the required PC ladder program as in the chapter example.
- ☐ Trace a number, if requested, through the computational system similar to the tracing carried out in figure 23–4.

3. The linear input of 0 to 80 volts is to be displayed on a 9999-maximum-count BCD output. Trace 32 volts through the system.

4. Repeat exercise 3, changing the output to a linear 0-to-21 volts. Trace 53 volts input through the system.

5. Two linear input signals of 0 to 4 volts are to be multiplied and the result put out on a linear output of 0 to 150 volts. Trace the numbers if the inputs are 2.85 and 3.45 volts.

6. Two BCD numbers are to be inputted. The first is to be divided by the second. The result is to be shown on an output BCD display. Trace the computation if A is 458 and B is 35.

7. There are three BCD inputs, A, B, and C. The output is to be A plus B minus C on a BCD display. Trace the computation for an A, B, and C of 425, 283, and 63, respectively.

8. There are two BCD inputs. If A exceeds 355, output F is to go on; if B exceeds 187, output G is to go on; if both exceed their listed numbers, output H is to go on; otherwise, no outputs are to be on.

Other Advanced PC Functions

24

At the end of this chapter, you will be able to

☐ Describe the hardware and software of the I/O module with its own internal computer

☐ Describe the following functions and program examples of each: Immediate Update; Transitional Functions; Continuous Select; Ascending Sort

☐ Describe the following functions: Transmit Print; First In/First Out and First In/Last Out; Program and Watchdog Timer Controls; Loop Process Control

INTRODUCTION

In addition to the PC functions described in the first 23 chapters, there are a number of other functions available. The availability of these other functions varies from manufacturer to manufacturer and model to model. These functions vary in name, as well. A given advanced function will have different designaions and formats.

This chapter describes nine of these commonly available functions. The first advanced function, the "smart" I/O module, involves hardware and software. The other eight functions are all software based. These advanced functions vary from straightforward, easily programmed functions to the complex Loop Control PID control function.

I/O MODULES WITH BUILT-IN COMPUTER

Some I/O modules, sometimes called "smart" modules, have built-in computers. They are auxiliary computers, and do not have the computing power of the main CPU. Figure 24–1 shows how these modules can be used. If immediate action is required that cannot wait for the CPU decision, the I/O module computer is used. For example, a pressure monitor signal, which indicates that an over-pressure is occurring, is sent into input port 03. The I/O computer monitor is programmed to deenergize out-

Figure 24-1
I/O Module with
Built-in
Computer

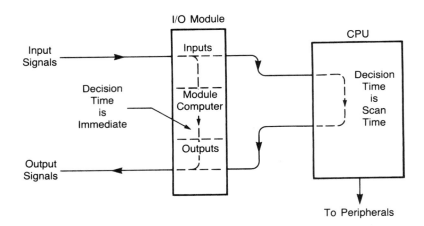

put terminal 07 immediately when input 03 shows over-pressure. Output terminal 07 is the process enabler, so the process is shut down.

All other non-critical logic analysis is done by the CPU. Note that *immediate* must be defined in milliseconds or microseconds. The time intervals of the process operation can affect its performance and must be considered.

IMMEDIATE UPDATE

The IMMEDIATE UPDATE (IU) function is used when you cannot wait until the end of a scan to update a portion of the ladder program; safe operation conditions, for example, may require it. Suppose scan time is 6 milliseconds and monitoring of a critical input must be updated every 1.5 milliseconds or sooner. The 1.5-millisecond updating can be accomplished by placing an IMMEDIATE UPDATE function for the critical input at 4 or 5 equally spaced points in the ladder diagram. If you insert 4 IU functions with equal spacing, you will have 5 updates, including the automatic one at the end. Six milliseconds divided by 5 updates is 1.20 milliseconds between updates, which is less than the required 1.5 milliseconds. Figure 24–2 shows a typical IU function that you could use when you do not have an I/O computer as described in the previous section.

The IU function is programmed using a pair of registers for action. The registers can be individual registers or OG or IG registers. When enabled, the status of the input-specified register is transferred to the output-

Figure 24-2
The IMMEDI-
ATE UPDATE
Function

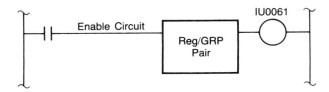

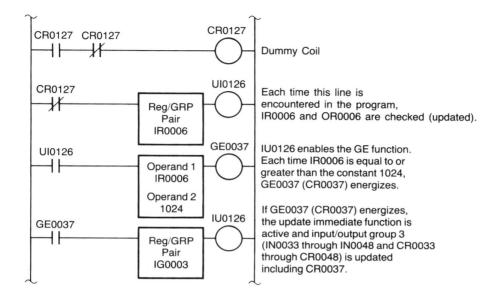

Figure 24-3
IMMEDIATE
UPDATE
Application

specified register. Individual logic device status must come in and go out from the specified registers.

Figure 24-3 shows how the function can be nested within associated ladder rungs. Following the 1.5-millisecond example, this updating could be placed at repeated intervals throughout the ladder diagram. See your PC manual to make sure that identical IU functions can be repeated without violating your PC's repetition rules. The IU coil number for the function is normally used for internal logic only, and goes on after the updating is completed.

The IU function could also be used for ladder logic sequence requirements. It might be necessary to update an output or logic coil before the end of the scan. Normal updating at the end of the scan could be out of the logic order needed for proper sequencing. This problem is rare, but could be a factor in some applications.

TRANSITIONAL FUNCTIONS

A TRANSITIONAL function is one that reacts and operates only once—on the first scan of the program. In some PCs you can type it in as a direct function, as shown in figure 24-4A. In other PCs it must be created as a "one shot" control, as shown in figure 24-4B.

An application of the TRANSITIONAL function might be the turning on of a light for 30 seconds to illuminate safety rules for the operator at the start of a new shift. The rules sign is turned off during the rest of the shift to enhance the initial effect. Figure 24-5 shows how the sign control could be accomplished.

Figure 24-4
The TRANSI-
TIONAL
Function

Transitional Output Instruction

A transitional output is an output that is on for the time it takes the processor to make one scan through the memory (it is turned off the next time it is scanned). For the output to turn on again, the input conditions controlling the output must make the transition from closed to open and back to closed.

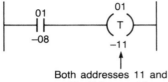

Both addresses 11 and 12 will be used for this transitional output.

Circuit Operation:
When contact 01-08 makes the transition from open to closed, output 01-11 will turn on. On the next scan, output 01-11 will turn off.

A

On Going One Shot

Off-Going One Shot

B

Figure 24-5
TRANSITION
Function
Application

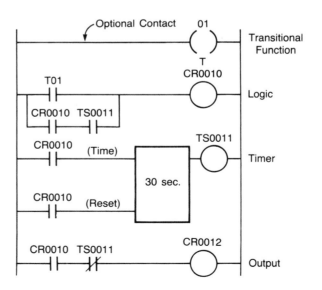

288

SELECT CONTINUOUSLY

If a new equipment operator was put in charge of the process in the IM-MEDIATE UPDATE example, you might want the sign to remain on for some time. It would be convenient to override the TRANSITIONAL function's one-time operation. The continuous override can be accomplished by using the SELECT CONTINUOUSLY (SC) function, as shown in figure 24–6. When on, the function keeps T–01 on continuously.

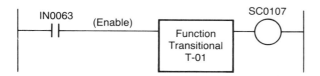

Figure 24–6
The SELECT CONTINU-OUSLY Function

ASCENDING SORT

Many advanced PCs have a SORT function that arranges a group of values into a low-to-high numerical listing order. The PC SORT operations are similar to those in other computers. A typical PC SORT is shown in figure 24–7.

To carry out the SORT, the function must know the series of registers, or addresses, where the unranked values are to be found. The function also has to know where to put the sorted and ranked values after they are rearranged in sequential order. An equal number of registers are required for input values and for receiving the resulting ranked values. In addition, some PCs require a third group of registers to be reserved. The third group, often called "scratch pad" registers, is used for intermediate calculating. Note also that more than one scan may be needed to complete the sort. An ASCENDING SORT function is shown in figure 24–8. The coil goes on after the last scan when the sort is completed.

For the function, *length* asks how many numbers are to be sorted; in this example, 53. *Input end* is the last input register used. If you used

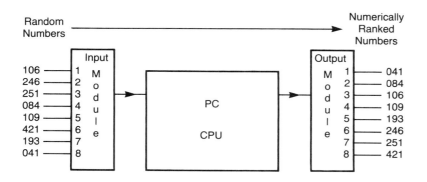

Figure 24–7
An ASCENDING SORT Problem

Figure 24-8
The
ASCENDING
SORT Function

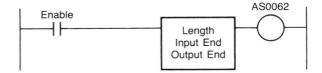

HR 0001 through HR 0053, insert HR 0053. For *Output end,* if you use OR 0150 through OR 0202, insert OR 0202. Some "scratch pad" registers might be required, which would have to be reserved also. See the PC operating manual to determine how many of these registers must be reserved. The function sorts when enabled is energized.

TRANSMIT PRINT

The TRANSMIT PRINT function, also called ASCII Print, stores a message in a group of registers or addresses in binary coded form. When the function is energized, the stored register bits are retrieved and sent sequentially to a designated output alphanumeric display, or to another peripheral or CPU. The ASCII code is the code most commonly used.

A typical application of the TRANSMIT PRINT function, as shown in figure 24-9, is the display of trouble messages; for instance, "Motor #7 is Overheated." The TRANSMIT PRINT function is inserted in the ladder diagram at a certain point in the process, to be enabled by an input from an overtemperature device in the motor.

Figure 24-9
The TRANSMIT
PRINT Function

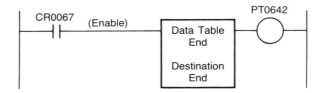

FIRST IN/FIRST OUT AND FIRST IN/LAST OUT

The FIRST IN/FIRST OUT (FI/FO) and FIRST IN/LAST OUT (FI/LO) functions are particularly useful in controlling inventory flow, among other uses given in application manuals. An example of the use of FI/FO is in programming operations for the day in the order of production. Suppose you must produce 125 different parts in one day. The codes for processing each different part would be entered in order. As production proceeds through the 125 parts, the operational codes are pulled out in order of production by FI/FO.

FI/LO might be used in quality control. For example, the defect codes for any defective parts are recorded in registers in order of defect occurrence. When pulling the data for analysis, you would use FI/LO to look at the last defect first—and progress back through the day's problems.

PROGRAM AND WATCHDOG TIMER CONTROLS

The PROGRAM and WATCHDOG TIMER CONTROL functions are used primarily in successive repetitive operations within a ladder program; for example, in successive approximations. A portion of the ladder diagram is repeated until the results of the repeated portion get within predetermined limits. Operational manuals show how to apply these functions.

LOOP CONTROL

Chapter 23 discussed analog control, which can be used to position valves and to set flows, levels, temperatures, and so on. A certain input signal value produces a specific output value or position. Chapter 23 examined open loop systems with no feedback corrections. With no corrective feedback, you have no assurance that the output is where it is supposed to be; you have only the assurance that a signal was sent. Chapter 23 did not cover situations in which the elapsed time taken for the output to reach the final point was a factor.

There are three factors regarding loop control still to be considered. First, a value set for an input does not reach the output immediately. Such factors as weight and speed of the process affect the time it takes for the output value to match the newly set input value. For instance, a valve weighing 7 ounces would track faster than a large valve weighing 350 pounds with equivalent drive mechanisms. Some compensating control might be required to make their movement speed and time identical.

Second, your first setting of the input might not give the output result you want. You might have to readjust the input repeatedly before the output result is correct, or sufficiently close to correct.

A third factor in analog processes is drift. After adjusting the process to obtain the desired output results, the process values and settings can drift away from their initial set points, and the initial output setting may no longer be the same at a later time. The input adjustment process must then be repeated periodically to preserve the required output.

Feedback control, sometimes called servo control, automatically compensates for these three error factors. With feedback, the output is monitored by a sensor that constantly indicates output position or status. The CPU compares the sensor status against the required status. If the signals differ, the input signal is appropriately modified to correct the output. Figure 24–10 shows a simple feedback loop system, in block diagram form, in which a 500-pound inertia wheel is to rotate. Its position is to match exactly the setting of the input dial. When the input setting signal and

Figure 24-10
Feedback Loop
System

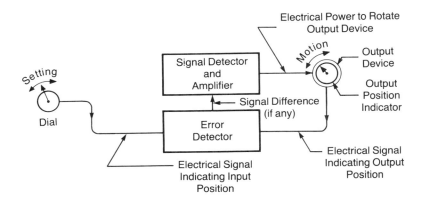

the output position indicator signal match, there is a zero-value differential error signal. No further output movement takes place. If the dial is then rotated to a new position, a difference-induced signal results, sending an electrical signal to the rotary motor, which moves the inertia wheel to its new, proper position.

Advanced PCs have a LOOP CONTROL function that is flexible and adaptable enough to be used where individual control packages were once required. The PC Loop Controller is a universal device that is usable in different processes and control applications.

Figure 24-11 shows the operation of an ideal analog system in which a large valve must be rotated from 0 to 108 degrees. The system has an amplification factor of 2 degrees per volt. When you apply 54 volts to the input for 3 seconds, the valve ideally goes instantaneously to 108 degrees.

The instantaneous movement shown in figure 24-11 could never happen because the valve needs some time, say, 4 seconds, to rotate 108 degrees. This more realistic but still somewhat ideal movement (which might still be too long) is shown in figure 24-12.

In actual industrial systems, the time-versus-position curves take other forms, as shown in figures 24-11 through 24-16. There are inherent control accuracy problems in each curve's situation: figure 24-13 shows the valve reaching incorrect positions; figure 24-14 shows an overshoot followed by decreasing oscillations; figure 24-15 illustrates an unacceptable length of time for reaching the final value; and figure 24-16 shows the valve continuously oscillating about its desired setting.

There is a LOOP CONTROL PC function called Proportional-Integral-Derivative (PID) that makes improved control of the valve movement possible. The bibliography in appendix D lists a number of reference books that detail the PID control theory. Figure 24-17 illustrates how the valve position curve would appear under PID control. The proper valve position is reached correctly, in a shorter time, and with a minimum of oscillation.

Each manufacturer explains its own Loop Control system in its users manual. Figures 24-18 and 24-19 display a portion of one manufacturer's

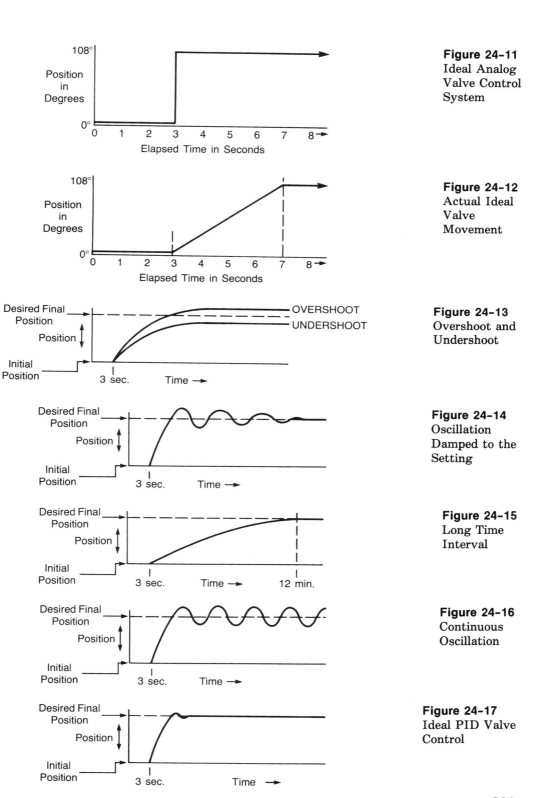

Figure 24-11
Ideal Analog
Valve Control
System

Figure 24-12
Actual Ideal
Valve
Movement

Figure 24-13
Overshoot and
Undershoot

Figure 24-14
Oscillation
Damped to the
Setting

Figure 24-15
Long Time
Interval

Figure 24-16
Continuous
Oscillation

Figure 24-17
Ideal PID Valve
Control

Figure 24-18
PID PC
Function

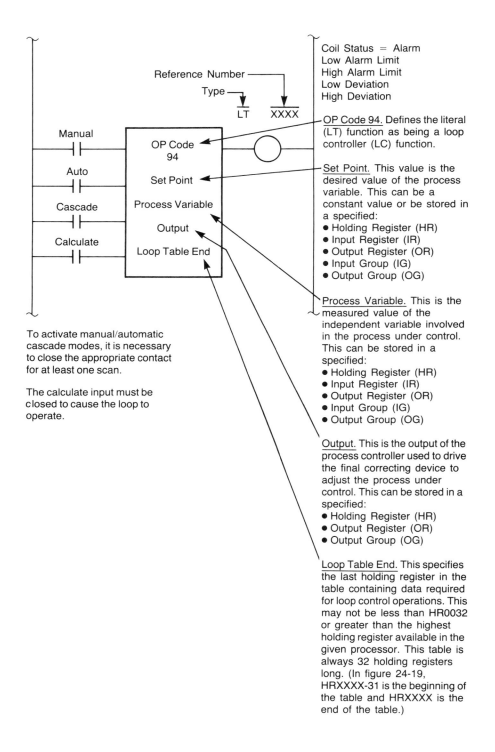

Reference Number

Type

LT   XXXX

Manual

Auto

Cascade

Calculate

OP Code
94

Set Point

Process Variable

Output

Loop Table End

Coil Status = Alarm
Low Alarm Limit
High Alarm Limit
Low Deviation
High Deviation

OP Code 94. Defines the literal (LT) function as being a loop controller (LC) function.

Set Point. This value is the desired value of the process variable. This can be a constant value or be stored in a specified:
• Holding Register (HR)
• Input Register (IR)
• Output Register (OR)
• Input Group (IG)
• Output Group (OG)

To activate manual/automatic cascade modes, it is necessary to close the appropriate contact for at least one scan.

The calculate input must be closed to cause the loop to operate.

Process Variable. This is the measured value of the independent variable involved in the process under control. This can be stored in a specified:
• Holding Register (HR)
• Input Register (IR)
• Output Register (OR)
• Input Group (IG)
• Output Group (OG)

Output. This is the output of the process controller used to drive the final correcting device to adjust the process under control. This can be stored in a specified:
• Holding Register (HR)
• Output Register (OR)
• Output Group (OG)

Loop Table End. This specifies the last holding register in the table containing data required for loop control operations. This may not be less than HR0032 or greater than the highest holding register available in the given processor. This table is always 32 holding registers long. (In figure 24-19, HRXXXX-31 is the beginning of the table and HRXXXX is the end of the table.)

294

Loop Table Register Positions	Loop Table Actual HR Assignment	Quantity		Value/Remarks
HRXXXX-31		Proportional Term ($\pm 32,767$)	C	
HRXXXX-30		Integral Term ($\pm 32,767$)	C	
HRXXXX-29		Derivative Term ($\pm 32,767$)	C	
HRXXXX-28		SP_n — Set Point This Sample	C	
HRXXXX-27		PV_n — Process Variable This Sample	C	
HRXXXX-26		Time Counter — Elapsed Sample Time	C	
HRXXXX-25		SP_{n-1} — Set Point Previous Sample	C	
HRXXXX-24		PV_{n-1} — Process Variable Previous Sample	C	
HRXXXX-23		E_{n-1} — Error Previous Sample	C	
HRXXXX-22		Bias (0 to Maximum Output)	C	
HRXXXX-21		RESERVED		FUTURE — DO NOT USE
HRXXXX-20		Configuration Input Word (See Below)	U	
HRXXXX-19		RESERVED		FUTURE — DO NOT USE
HRXXXX-18		RESERVED		FUTURE — DO NOT USE
HRXXXX-17		Integral Sum ($\pm 32,767$)	C	
HRXXXX-16		E_n — Error This Sample	C	
HRXXXX-15		T_d — Derivative Time (0 — 327.67 Min.)	U	
HRXXXX-14		T_i — Integral Time (0 — 327.67 Min.)	U	
HRXXXX-13		T_s — Sample Time (0 — 3276.7 Sec.)	U	
HRXXXX-12		K_c — Proportional Gain (.01 — 99.99)	U	
HRXXXX-11		Inner Loop Pointer (Loop Table End)	U	
HRXXXX-10		Outer Loop Pointer (Loop Table End)	U	
HRXXXX-9		Alarm Deadband (0 — Max PV)	U	
HRXXXX-8		Batch Unit Preload (0 — Max Output)	U	
HRXXXX-7		Batch Unit Hi Limit (0 — Max Output)	U	
HRXXXX-6		Neg. Slew Limit (Max — Δ Output/Sample)	U	
HRXXXX-5		Pos. Slew Limit (Max + Δ Output/Sample)	U	
HRXXXX-4		Low Deviation Alarm Limit (0 — Max PV)	U	
HRXXXX-3		High Deviation Alarm Limit (0 — Max PV)	U	
HRXXXX-2		Low Alarm Limit (0 — Max PV)	U	
HRXXXX-1		High Alarm Limit (0 — Max PV)	U	
HRXXXX		Output Status Word	C	

C = Calculated by Processor
U = User-Entered

Configuration Input Word (HRXXXX-20)

16 15 14 13   12 11 10 9   8 7 6 5   4 3 2 1

Bit Number	Definition	Status	Bit Number	Definition	Status
1	1 = Proportional Mode Selected		9	1 = Derivative on PV Selected 0 = Derivative on Error Selected	
2	1 = Integral Mode Selected		10	1 = Batch Unit Selected	
3	1 = Derivative Mode Selected		11	RESERVED FOR CONTROLLER USE	
4	1 = Deviation Alarms Selected		12	0 = Anti Reset Windup When Slew Limit Occurs	
5	1 = Error Deadband Selected		13	RESERVED FOR FUTURE USE	
6	1 = Error Squared Control Selected		14	RESERVED FOR FUTURE USE	
7	1 = Slew Limiting Selected		15	RESERVED FOR FUTURE USE	
8	1 = Reverse Action Selected 0 = Direct Action Selected		16	RESERVED FOR FUTURE USE	

Figure 24–19
PID PC Function Instruction Set

instructions. As you can see, the whole setup is quite complicated and requires considerable programming time.

EXERCISES

1. A chemical process has a critical safety point of 87 °C to be checked. If the temperature rises above 87 °C, an output signal must shut down the process within 0.75 milliseconds. The critical point check input, IN 0072, goes off instantaneously if the 87 °C is reached or exceeded. Output OR 0123, the feed signal terminal to the shutdown device, is normally energized during operation. The PC scan time is 4.75 milliseconds. Create an IMMEDIATE UPDATE PC system for the required protection. Assume that the same IU ladder rung can be added at periodic points in the ladder diagram.

2. At the beginning of the day's operation, a sign listing safety precautions is to appear for operator review. After two minutes, the sign is to go off for the day. Show how a TRANSITIONAL function and a timer could be used to accomplish this.

3. For a new operator, the safety listing sign for exercise 2 is to be on all day. Show how a CONTINUOUS SELECT function can be used to override the transitional function.

4. A group of 17 numbers with random values between 1 and 100 are input into a PC system and placed in holding registers 62 through 78. Show how the numbers could be sorted by the PC and arranged in numerical sequence. The resulting sorted numbers are then output from output registers 201 through 217. Reserve the required intermediate, or "scratch pad," registers if needed.

Glossary

Abort   The action of terminating the progressive operation of a program or process operation.

AC   Alternating Current. Electrical current normally alternating 60 times per second.

Address   A specific location in a computer memory. Represented by a number, label, or name.

Analog Signal   A continuous value between two limits. Can represent position, voltage, angle, or any electrical signal with a varying value.

Analog Input Module   A PC module with terminals capable of receiving a continuous, varying, electrical value from an outside device or process.

Analog Output Module   A PC module with terminals capable of furnishing a continuous, varying output voltage to an output device.

Analog-to-Digital Converter   A circuit for converting a varying analog signal to a corresponding representative binary number.

AND (logical)   A logic gate whose output is on only if all of its inputs are on.

Arithmetic Logic Unit (ALU)   A computer CPU subsystem that can perform arithmetic and logic gate operations.

Arithmetic Capability   The ability of a computer to perform math functions.

ASCII   American Standard Code for Information Interchange. A 7-bit code for representing letters, numbers, and symbols appearing in written material.

Baud   A rate of data transmission. Its rate is equal to the number of code elements per second that are transmitted.

BCD   Binary Coded Decimal is a system in which each decimal digit from 0 through 9 is represented by a pattern of four binary bits.

Binary   A system of counting using only 0 and 1.

Binary Number System   Sometimes called *base two*, a numbering system using the digits of 1 and 0 only.

Binary Word   A group of bits usually found in a single register or address location.

Bit   A single binary digit. Can have a value of 1 or 0. From *Binary digIT*.

Bit Pick   A PC system of choosing one bit in a register to determine a status of a PC input or output. A bit of 1 means on and a 0 means off.

Boolean Algebra   A shorthand notation that expresses logic functions in equation-type expressions.

Boolean Equation   Expresses the relations between logic functions in an equation written in Boolean algebra form.

Branch   A parallel logic path within a user program rung.

Buffer A temporary storage area in a computer used for intermediate storage of data. Typically receives data at one rate and then outputs the data at a different rate or in a different form.

Bus One or more conductors for transmitting data between destinations.

Byte A sequence of binary digits usually operated upon as a unit. The exact number of digits will vary with different systems, but is normally 4, 8, 16, or 32.

Cascading Placing two or more functions of the same kind in sequence. The purpose of cascading is to extend the number of operational steps beyond that of one function.

Cassette Recorder/Player For programmable controllers, a device that can transfer information between PC memory and magnetic tape. When recording, it makes a permanent record on tape of a program or data from a processor memory. In the playback mode, the cassette recorder enters the previously recorded program or data from the tape into processor memory.

Central Processing Unit (CPU) The central control unit of a programmable controller logic system.

Chaining See cascading.

Character One symbol of a set of basic symbols, such as a decimal number, a punctuation mark, or a letter.

Chip Another name for an integrated circuit. A tiny piece of layered semiconductor material mounted in a small case with terminals. Contains a large number of transistors, resistors, and capacitors in miniature.

Clear A command to remove data from one or more memory locations. Normally sets the memory location value to zero.

Clock A circuit that generates timed pulses to synchronize the timing of computer operations.

Coaxial Cable A special type of tubular computer-connecting cable having two electrical paths. One is a wire in the center; the other is circular, braided material outside of tubular insulation around the center wire. May have a second, outer, braided cable for electrical shielding.

Code A system of symbols or bits for representing data, ideas, or characters.

Coil Represents the output of a programmable controller. In the output devices it is the electrical coil that, when energized, changes the status of its corresponding contacts.

COMPARE Function A programmable controller function that compares two numbers to see if they meet or do not meet the specified criterion.

Computer Interface A device that communicates between various computers or computer modules.

Contact A part in a relay with two terminals. In PCs, it is in a conducting or non-conducting state, depending on its corresponding coil's status and the coil's initial status, whether normally open or normally closed.

Control Relay See relay.

Counter A device for counting input pulses or events. Its output changes status when the preset number of counts is reached.

CPU See Central Processing Unit.

Cross Reference In ladder diagrams, letters or numbers to the right of coils or functions. The letters or numbers indicate on what other ladder lines contacts of the coil or function are located. Normally closed (NC) contacts are distinguished from normally open contacts (NO) through the use of an asterisk (*) or by underlining.

CRT Cathode Ray Tube. An electronic viewing tube on which PC data is displayed in alphanumeric or graphic form.

Cursor An indicator that shows where the computer's action pointer is located. Any data entered through the keyboard will occur where the cursor appears on the monitor screen in the EDIT mode.

Data Transfer A computer operation that moves data from one address or register to another.

DC Direct current. Electrical current flowing continuously in the same direction, usually at a fixed rate or value.

Debug Correcting mistakes in a program through various forms of analysis.

Decimal Number System The base-ten system of counting. Numbers are 0 through 9.

Diagnostic Program A computer program used to analyze faults in another computer program or in a system's operation.

Digital A system of discrete states, on or off, high or low, or 1 or 0.

Digital-to-Analog Converter An electrical circuit that converts binary bits to a representative, continuous, analog signal.

Digital Gate A device that analyzes the digital states of its inputs and puts out an appropriate output state.

Dip Switch A group of small, in-line, on-off switches. From *D*ual *I*nline *P*ackage.

Discrete Having the characteristic of being on or off.

Discrete Input Module A PC module which processes input status information having two states only—high or low.

Discrete Output Module A PC module that puts out only two states—on or off.

Disk Drive A device that records or reads data from a rapidly scanned flat disk.

Diskette The flat, flexible disk on which a disk drive writes and reads.

Documentation A logical, orderly, recorded or written document containing software or data listings.

Double Precision The system of using two addresses or registers to display a number too large for one address or register. Allows the display of more significant figures since twice as many bits are used.

Down Counter A counter that starts from a specified number and increments downward to zero.

Down Load Loading data from a master listing to a readout or another position in a computer system.

Drum Switch Synonymous with *sequencer*. Normally mechanical in nature, it operates through a multiple sequence of simultaneous on-off states.

Dump Recording the contents of a computer memory or of computer data on a tape, disk, etc. Normally done for backup in case of computer malfunction or the loss or distortion of computer data.

EBCDIC Extended Binary Code Decimal Interchange Code. A code similar to ASCII for representing letters, numbers and symbols, except that it uses 8 bits instead of 7.

EEPROM Electrically Erasable Programmable Read Only Memory. A programmable IC chip. The program, after use, can be erased (all bits reset to zero) by applying an electrical current to two of its terminals.

Element A part of a PC program. Coils, contacts, timers, and the ADD function are examples of PC program elements.

Enable To allow a function to operate by energizing a PC ladder line. If not enabled, the PC function will not be active.

Encoder A rotating device that transmits a coded feedback signal indicating its various positions.

EPROM Erasable Programmable Read Only Memory. Same as EEPROM, except that resetting is accomplished by exposing a small section under a "window" to ultraviolet light.

Examine Off An instruction that is true only if the examined or addressed bit is off or 0.

Examine On An instruction that is true only if the examined or addressed bit is on or 1.

EXOR *EX*clusive *OR* gate. A digital gate in which the output is on only when one of its two inputs, not both, are on.

Fail-Safe A control situation that discontinues the operation of a process when the control power source fails. A non-fail-safe operation requires the application of control power to turn it off.

False Prescribed conditions are not met and the logic is disabled or remains disabled.

Feedback In analog systems, a correcting signal received from the output or an output monitor. The correcting signal is fed to the controller for process correction.

File A set of logically arranged data. A file is stored in various computer locations, normally consecutively.

Floppy Disk A recording disk used with a computer disk drive for recording data. The disk is flexible, not rigid.

FORCE Function A keyboard function used to turn elements of a PC ladder diagram on and off. It overrides the input function status received through the input module.

Format Refers to the language arrangement and layout for a given type of PC system.

Gate Electronically, a device that makes logic decisions, depending on its input statuses; the gate's output is turned on or off, accordingly.

Gray Code A special digital code similar to the binary code, except that only one of its bits changes status when going sequentially from one number to the next.

Ground An electrical connection made for safety from a unit case to ground potential. Also sometimes defined as an undesirable connection from an electrical device's electrical system to its case, earth electrical potential, or to the computer's zero-potential level.

Hand-Held   A small, portable, programming keyboard, usually with a small LCD window on which portions of the entire ladder diagram may be displayed.

Hard Copy   A printed copy of data, programs, etc.

Hard Disk   An inflexible recording disk used with a computer disk drive.

Hardware   In contrast to software, the mechanical, electrical, and electronic parts of a computer.

Hex   An abbreviation for hexadecimal.

Hexadecimal   A numbering system with four binary bits. Represents 0 through 15 in the decimal system by using the digits 0 through 9 and the letters A through F.

High   A status representation of on, 1, or true.

Holding Register   A type of address location in the CPU for symbol or logic storage.

Host Computer   A main computer that controls other computers, PCs, or computer peripherals.

Hydraulic   A system of control using a fluid.

I/O   An abbreviation for input/output.

I/O Module   The electronic assembly of a PC system that interfaces between the PC CPU and the "outside world."

I/O Rack   See rack.

I/O Update Time   The time interval in milliseconds that it takes for a PC to update the statuses of all input and output modules.

Input Devices   Devices connected to the PC input modules for sending status information. Switches, limit switches, pushbuttons, and electrical potentiometers are examples of input devices.

Input Group Register   A single register into which the CPU records the statuses of a group of 8 or 16 input registers.

Input Module   An electrical unit or circuit used to connect input devices electrically to the PC. A module sends coded signals to the CPU indicating the status of each input.

Input Register   A PC register (address) associated with input devices.

Integrated Circuit   See Chip.

Interfacing   Connecting a PC to outside inputs or outputs, or connecting different computers to each other.

Interlock   A system for preventing one element or device from turning on while another device is on. May be applied between more than two elements or devices.

Jog   A state of being momentarily on or in motion. In controls, the momentary on state is caused by depressing a spring-return switch or push button. When the switch or push button is released, the device returns to the off state and is not sealed on.

Jump   A command in a computer program that causes the sequence to go to, or branch to a specified point, though not necessarily the next point in the program sequence.

Keyboard The alphanumeric keypad on which the user types instructions to the PC.

Label The means of identification of registers, addresses, contacts and coils—normally in letters, numbers, or alphanumeric.

Ladder Diagram A system of successive horizontal lines with symbols representing the logic operation of a control system. The symbols are drawn in relay-logic or PC-logic form. The control contacts are to the left and coils and functions are on the right.

Language A group of letters and symbols used for intercommunication between persons, computers, or persons and computers.

Latch An electronic or mechanical device that causes an energized coil to remain on after its input signal is turned off.

Latching Relay A relay with a latching-type operation and two inputs, on and off.

LCD Liquid Crystal Display. A type of small, numerical indicating segments that reflect light and, in combination, provide a visual display.

LED Light-Emitting Diode. A type of small light used in combinations to give a visual display by emitting light.

Limit Switch A mechanical device that turns a built-in electrical switch on or off. The electrical switch is actuated by depressing its protruding arm.

Loop Control A control of a process or machine that uses feedback. An output status indicator modifies the input signal effect on the process control.

Low A state of being off, 0, or false.

Magnetic Tape A thin, plastic tape covered with magnetic particles. The tape stores information by becoming magnetized at specific points as it passes through a fixed location. The stored information may be read from the tape on a subsequent pass-through.

Matrix An arrangement of data or circuit elements in an X-Y, two dimensional array. The matrix may have any dimensions, but is often 8 by 8 or 16 by 16 in a PC.

Mechanical Drum Programmer See drum switch.

Memory In a PC, the group of addresses or registers where information and programs are stored. Storage may be permanent or temporary and erasable.

Menu A list of programming choices shown on a PC screen.

Microprocessor A computer on a chip containing functions normally found on many different chips. It has all of the capabilities of the digital computer, such as ALU, memory, logic, and registers.

Microsecond One thousandth of a second.

Millisecond One millionth of a second.

Mnemonic Codes A short code for a function, usually an abbreviation or combination of key letters for easy recognition. For example, SB for SUBTRACT, EQ for EQUAL TO, and MCR for MASTER CONTROL RELAY.

Mode The functional form in which a computer is operating; for example, run, or program.

Module An electronic functional device that may be attached to or plugged into a bus. The bus is connected electrically/electronically to other modules and, more often, to a main computer.

Monitor A mode of operation in which information is displayed. A screenlike device for displaying the program, the operational status of the PC, or the status of the process the PC is controlling.

NAND A digital gate whose output is off only when all of its inputs are on.

Negative Logic In digital logic, a system in which the high or on state is more electrically negative than the low or off state.

NEMA Standards A set of industry standards of dimensions, specifications, and performance parameters published by the National Electrical Manufacturers Association.

Nesting In ladder diagrams, locating a series of contacts logically within another series of contacts.

Network A number of interconnected logic devices.

Node A common electrical or logic point with two or more points of the circuit or diagram connected to it.

Non-Retentive Describes a PC logic device that loses its count of increments when the input goes off or low.

NOR A digital gate whose output is off when either one or more of its inputs are on.

Normally closed Contact (NC) A contact that is conductive when its operating coil is not energized.

Normally Open Contact (NO) A contact that is non-conductive when its operating coil is not energized.

NOT A digital inverter gate. On translates to off, and off translates to on through the gate.

Octal A numbering system using three binary bits equivalent to a decimal 8. Numbers used are 0 through 7.

Off-Delay Timer A timer that initiates an action at a specified time after another action ceases.

One Shot An action that takes place once per initiation. Once started, the action lasts for its specified period regardless of further status changes of its initiating input.

Operand A number used in an arithmetic operation as an input.

Optical Isolation Electronic isolation of two parts of a circuit by using a small light beam between the two stages. One stage produces a light beam of appropriate varying intensity and the other receives and decodes the varying light's pattern.

OR A digital gate that is on if any one or more of its inputs are on.

Output An electrical signal from a PC used to control a process device under the PC's control.

Output Devices Devices connected to the output modules to receive status information. Relay coils, fans, lights, and motor starter coils are examples of output devices.

Output Group Register A PC register (address) that can control multiple outputs, on or off, by the status of its individual bits. Usually controls 8 or 16 outputs.

Output Module   An electrical unit or circuit used to connect the PC to outside devices which are to be turned on or off.

Output Register   A PC register (address) associated with output devices.

Parallel Circuit   An electrical or control circuit in which the opposite ends of two or more components or elements are each connected to the same nodes. A parallel circuit may make up the whole circuit or a portion of an overall larger series parallel circuit.

Parallel Transmission   A computer operation in which two or more bits of information are transmitted simultaneously.

PC   Programmable Controller.

Peripherals   In computer systems, the devices connected to or controlled by the computer's central processor.

Pick   A method of selecting the state of a specific, single, register bit.

PID   Proportional-Integral-Derivative. A sophisticated analog control system for accurately and speedily controlling output parameters.

Pilot Run   A pre-production run of a small amount of product to work out any "bugs" in the process or program before going into full production.

Pneumatic   A system run by air pressure.

Port   In computers, a point of connection to a peripheral input or output device.

Positive Logic   In digital logic, a system in which the high or on state voltage is more positive than the low or off state voltage.

Power Supply   For computers, the device that converts line power, usually 120 volts AC, to the power type required by the computer, which can be various values of low-voltage DC.

Program   A logical sequence of computer steps carried out sequentially by the computer.

Programmer   A keyboard or other device used to enter a program into the computer. Can also monitor, edit, control, and modify the program.

Programmable Controller (PC)   A user-friendly computer that carries out control functions of many types and levels of complexity.

Protected Memory   Instructions or data stored in a computer memory that cannot be erased or altered.

Proximity Device   A non-contact, input-indicating device for detecting the presence of an object associated with a process. May be discrete or varying-value analog, depending on the process being controlled.

Rack   A mechanical channel or chassis on which PC input and output modules are mounted. May also include wiring channels and connectors.

Rated Voltage   The electrical voltage required for proper operation of a PC. Usually has allowable upper- and lower-range values. Example: $+5.0$ volts $+/- 0.2$ volts.

RAM   Random Access Memory chip. A chip which has read and write capabilities.

Read/Write Memory   A computer memory that can receive and store (read) information, and can be accessed for retrieval of stored information (write). Stored information can be erased or replaced. Writing out information does not change the stored information that has been accessed; the stored information is duplicated in the write destination.

Register   A location in a PC memory for storing information, usually in bit form. Essentially, a specified address.

Relay   A device actuated by a voltage or signal. The actuation changes the status of discrete mechanical or electronic contacts associated with the device.

Reliability   The ability of a device to perform its function correctly over a period of time or through a number of actuations or operations. Can be expressed as a decimal or a percentage.

Retentive Timer (or Counter)   A timer with two inputs. One is enable/reset and the other actuates the timing cycle. If the timing cycle is interrupted during its interval, the accumulated time is retained. When the input is reclosed, the timer starts at its retained time. The time may be reset to its initialized value only when the enable/reset input is deenergized. A retentive counter operates similarly.

ROM   Read Only Memory. An integrated circuit chip with unalterable information.

RS-232C   A standard type of computer interconnecting wiring system. Set by the Electronics Institute of America (EIA).

Rung   The ladder PC system which controls one output. May be on more than one horizontal control line.

Scan   The sequential operation of a PC that goes through the ladder diagram from top to bottom and updates all of the outputs according to input statuses. The scan usually takes place from left to right on each rung. Scan time is in microseconds for each scan and is repeated continuously in normal PC operation.

Scan Time   The time required for one complete sweep through the PC's entire ladder program.

Schematic   An electrical diagram symbolically showing components and their wiring schemes. It normally is in logic form, not in wiring connection diagram form.

SCR   Silicon Controller Rectifier. A solid-state thyristor device used to control DC power levels. Control of the levels is accomplished by varying the on or firing time within each AC pulse by pulses to the SCR gate terminal. Each single SCR operates only on one half of the sine wave input.

Scratch Pad Memory   A small group of registers or addresses used for intermediate computer calculation operations.

Seal   A contact paralleling an input device contact. The seal keeps the output on when the input device is turned off, since the parallel contact (seal) is controlled and turned on by the output.

Sensor   A PC input device that senses the process condition. Its status is fed to the PC through an input module.

Serial Operation   A system of transferring data sequentially rather than simultaneously.

Sequence   The order in which events take place.

Sequencer   A control system that sequences multiple devices through a fixed series of on-off statuses. See Drum Switch.

Series Circuit   A circuit in which elements are connected end to end. May be the whole circuit or a portion of an overall series parallel circuit.

Servomechanism   A control system that uses feedback for accuracy and process correction.

Significant Digit   A digit used to give the accuracy required to a number or an operation.

Simulator   An external device used with a PC that represents the process to be controlled. It would contain input switches and signal devices, and discrete and analog output indicators. It is not as complicated or as costly as the process it represents.

Software   The programs that control a computer.

Solenoid   A magnetic coil that changes the on or off status of an output device. The change is accomplished by the movement of an iron plunger, which may be a spring-return type.

Solid State   Made of semiconductor material. May be transistors, thyristors, or complete circuits in the form of integrated circuits.

Status   The condition of a device, usually on or off.

Thumbwheel Switch   A series of small, adjacent numbered (0 through 9) rotary wheels that may be set to a given number. Their settings may be inputted to a PC for process control.

Timer   A device for monitoring or determining times. It begins timing at one event and causes another event to occur at the end of the specified time.

Toggle Switch   A small electrical switch with an extended lever for actuation. Usually panel mounted.

Triac   A solid-state thyristor device capable of being switched on at a given point in an AC voltage cycle. Works on both the positive and negative portion of the input sine wave. Essentially a two-way SCR.

True   Prescribed conditions are met and the logic is enabled or remains enabled.

Truth Table   A yes/no (1/0) matrix indicating the status of one function and how it depends on the status of one or more other functions.

TTL   Transistor-Transistor Logic. A type of logic on a chip that includes multiple transistors and gates.

Unlatch Instruction   A PC command that turns a function off and keeps it off, overriding any other subsequent instruction to turn it on.

Up Counter   An event counter that starts from 0 and increments up to the preset value.

User Friendly   A term indicating that a PC program operation can be run by a person who, with minimal training or instruction, can proceed through the program by following sequential instructions appearing on the screen.

Value   The number or symbol located in the position specified.

Volatile Memory   A memory of values or status that is lost when power is turned off. Memory locations are usually reset to 0 as a result.

Watchdog Timer   A timer for monitoring proper circuit operation. If a monitored process interval is not met, the timer puts out a signal used to terminate or shut down a process.

Word   A group of bits used to represent a number or symbol.

Write   For PCs, the insertion of information into an address or register.

Operational Simulation and Monitoring

An external simulator is a device used for simulating the operation of a programmable controller to train users and test program sequences. Simulator panels with switches for inputs and lights for outputs are available from some manufacturers. In other cases you may wish to build your own discrete logic simulator, or an analog simulator for those situations involving analog (variable) PC values. For an analog simulator, variable input voltage must be available from a potentiometer or BCD device, and analog output indicators must be variable value indicators, not just indicating lights.

Using a PC system for training is difficult without a simulator, because you must continually move shorting wires around to simulate input switch closure, and you must move a voltmeter around to observe outputs being on or off.

Some training and simulation situations have specific applications for which a simulator with actual input and output devices is desirable. For example, simulation of a PC used to control a process with pneumatic cylinders would need actual pneumatic cylinders for the output, and, for the input switches, sensors that work the same as those in an actual process. Required elapsed time would then be included in the simulator operation.

PC DISCRETE SIMULATORS

A discrete simulator is available from some manufacturers, or you can build your own I/O panel. A typical wiring scheme of a panel built in-house is shown in figure APB-1. An advantage of the panel shown is that it has both momentary and permanent contacts, both of which are found in industrial processes.

There are three precautions to be considered in building the panel: first, note that the input common connection and the output common connection are not the same polarity, must be separate, and cannot be cross connected; second, parallel resistors are needed with high-resistance neon bulbs because off-state PC output module leakage current can keep the bulbs glowing; and third, a solid, separate safety ground wire must be run from building ground and connected to the metal chassis and cabinets.

Figure APB-1
PC Discrete
Simulator

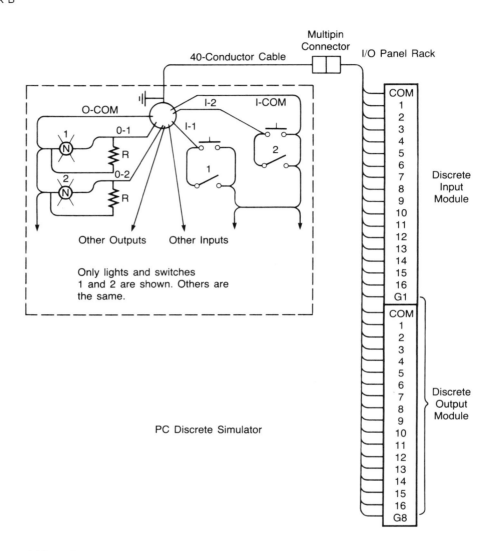

PC Discrete Simulator

ANALOG SIMULATORS

An analog simulator is shown in figure APB-2. The previous panels shown were for on-off, discrete values. When inputs and outputs are variable or analog, variable inputs and variable output indicators are used. Inputs may be BCD or a variable voltage device. Outputs may be a digital readout or an electrical analog device such as a meter.

Many analog simulators involve one input and one output only. The simulator should have two or more inputs to ensure the student understands the concept; otherwise, you might as well connect the input dial

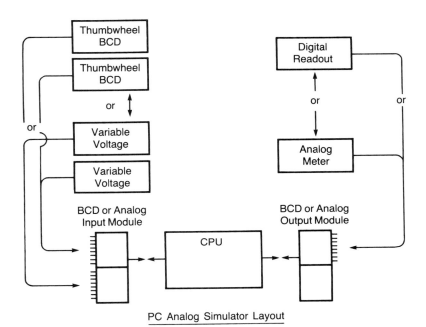

PC Analog Simulator Layout

to the output device directly, because you get the same effect whether you go through the PC or not. With two inputs as shown, actual CPU programming can be carried out for training and process simulation as discussed in chapter 23.

Commonly Used Circuit Symbols

I-Inputs

Limit Switch

Normally open contact	
Normally open contact held closed	
Normally closed contact	
Normally closed held open contact	

Toggle Switch

NO	NC
Toggle switch spring return	Thermocouple

Motor Centrifugal Switch

Speed (plugging)	Anti-plug
F F	F
R	R

General Switches

Float	NO	
	NC	
Thermal	NO	
	NC	
Pressure	NO	
	NC	
Foot	NO	
	NC	
Liquid Level	NO	
	NC	
Flow	NO	
	NC	
Proximity	NO	
	NC	

Push Buttons

Momentary Contact					Maintained Contact		Illuminated
Single circuit		Double circuit	Mushroom head	Wobble stick	Two single ckt.	One double ckt.	
NO	NC	NO & NC					

Selector

2 Position	3 Position	2 Pos. Sel. Push Button

2 Position:

	J	K
A1	1	
A2		1

1−contact closed

3 Position:

	J	K	L
A1	1		
A2			1

1−contact closed

2 Pos. Sel. Push Button:

Contacts	Selector Position			
	A		B	
	Button		Button	
	Free	Depres'd	Free	Depres'd
1−2	1			
3−4		1	1	1

1−contact closed

General Multiple Switch Convention

SPST NO		SPST NC		SPDT		Terms
single break	double break	single break	double break	single break	double break	SPST − single pole single throw
						SPDT − single pole double throw
DPST, 2 NO		DPST, 2 NC		DPDT		DPST − double pole single throw
single break	double break	single break	double break	single break	double break	DPDT − double pole double throw

KEY:
NO = Normally open
NC = Normally closed

II-Outputs

Timer	Contactor	Solenoid	Relay Coil
—(TR)—	—(M)—	—o—◊—o—	—(CR)—

Annunciator	Bell	Buzzer	Horn Siren, etc.	Meter	Pilot Lights		
				Indicate type by letter	Indicate color by letter		
					Non push-to-test	Push-to-test	
◇				VM AM	(A)	(R)	

AC Motors				DC Motors			
Single phase	3 phase squirrel cage	2 phase 4 wire	Wound rotor	Armature	Shunt field	Series field	Comm. or compens. field
◯	◯	◯	◯	◯	⌒⌒⌒ (show 4 loops)	⌒⌒ (show 3 loops)	⌒ (show 2 loops)

Resistor-fixed value

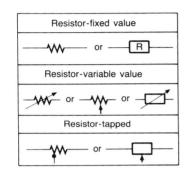

Resistor-fixed value
—ᗯᐧ— or —[R]—
Resistor-variable value
—ᗯ— or —ᗯ— or —
Resistor-tapped
—ᗯ— or —[ ]—

Capacitor

Capacitor	
fixed	variable
—)|—	—)|—
Inductors-fixed	
Air core	Iron core
ᴗᴗᴗᴗ	≡ᴗᴗ
Mechanical interlock	
– – – – – ∧ – – – – –	

Battery / AC Supply

Battery	AC Supply
+‖|‖|‖—	—(∿)—

Normally open relay contact	Normally closed relay contact
—| |—	—|/|—

Transformers

Auto	Iron core	Air core	Current	Dual voltage	Multiple output
⌇	⊐⊏	⌣⌣	⌒⌒	⌣⌣∧⌣⌣	⌣⌣⌣

Rectifier

Single phase half wave	Single phase full wave	Single phase bridge
▸|	AC …DC	DC ◇ DC, AC/AC

Symbols for static switching control devices

Static switching control is a method of switching electrical circuits without the use of contacts, primarily by solid state devices. Use the symbols shown in table above except enclosed in a diamond:

Timer

Function	Part	Symbol
On: Delay retards relay-contact action for predetermined time after coil is energized.	Normally open timed closing (NOTC) contact	⟋°
	Normally closed timed opening (NCTO) contact	⟍ᵒ
	Normally open timed open (NOTO) contact	⟍°
Off: Delay retards relay-contact action for predetermined time after coil is de-energized.	Normally closed timed closed (NOTC) contact	⟍ᵒ
	No instantaneous contact	—| |—
	No instantaneous contact	—|/|—

313

IV - Electrical Protection Devices

Fuse	Circuit Breaker		Overload Relays	
	air	oil	thermal	magnetic
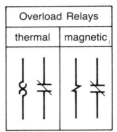				

Disconnect	Circuit interrupter	Circuit breaker w/thermal O.L.	Circuit breaker w/magnetic O.L.	Circuit breaker w/thermal and magnetic O.L.

V - Wiring Diagram Conventions

Not connected	Connected	Power	Control	Wiring terminal
				○
				Ground

Bibliography

PROGRAMMABLE CONTROLLERS

Cox, Richard D. *Technician's Guide to Programmable Controllers*. Albany, N. Y.: Delmar Publishers, 1984.

Jones, C. T., and L. A. Bryan. *Programmable Controllers—Concepts and Applications*. Atlanta: IPC/ASTEC Publishers, 1983.

Kissell, Thomas E. *Understanding and Using Programmable Controllers*. Englewood Cliffs, N. J.: Prentice Hall, 1986.

Programmable Controls—The User Magazine. Bimonthly Periodical: ISA Services, Inc., 67 Alexander Drive, P.O. Box 12277, Research Triangle Park, No. Caro. 27709.

CONTROLS

Bateson, Robert. *Introduction to Control System Technology*. Columbus: Merrill Publishing Company, 1980.

Herman, Stephen L., and Walter N. Alerich. *Industrial Motor Control*. Albany, N. Y.: Delmar Publishers, 1985.

Rexford, Kenneth. *Electric Controls for Machines*. Albany, N. Y.: Delmar Publishers, 1987.

Rockis, Gary, and Glen Mazur. *Electric Motor Controls*. Alsip, Ill.: American Technical Publishers, 1982.

Wiring Diagrams Bulletin, #SM304, Milwaukee: Square D Company.

INDUSTRIAL ELECTRONICS

Chute, George M., and Robert D. Chute. *Electronics in Industry*. New York: McGraw-Hill, 1979.

Moloney, Timothy J. *Industrial Solid State Electronics—Devices and Systems*. Englewood Cliffs, N. J.: Prentice Hall, 1986.

DIGITAL ELECTRONICS

Floyd, Thomas. *Digital Fundamentals*, third edition. Columbus: Merrill Publishing, 1986.

INDICATORS AND SENSORS

Honeycutt, Richard D. *Electromechanical Devices.* Englewood Cliffs, N. J.: Prentice Hall, 1986.

Seippel, Robert G. *Transducers, Sensors, and Detectors.* Reston, Virginia: Reston Publishing Company, 1983.

Index